Modeling in Medical
Decision Making

STATISTICS IN PRACTICE

Founding Editor

Vic Barnett
Nottingham Trent University, UK

Statistics in Practice is an important international series of texts which provide detailed coverage of statistical concepts, methods and worked case studies in specific fields of investigation and study.

With sound motivation and many worked practical examples, the books show in down-to-earth terms how to select and use an appropriate range of statistical techniques in a particular practical field within each title's special topic area.

The books meet the need for statistical support required by professionals and research workers across a range of employment fields and research environments. Subject areas covered include medicine and pharmaceutics; industry, finance and commerce; public services; the earth and environmental sciences, and so on.

The books also provide support to students studying statistical courses applied to the above areas. The demand for graduates to be equipped for the work environment has led to such courses becoming increasingly prevalent at universities and colleges.

It is our aim to present judiciously chosen and well-written workbooks to meet everyday practical needs. Feedback of views from readers will be most valuable to monitor the success of this aim.

A complete list of titles in this series appears at the end of the volume.

Modeling in Medical Decision Making

A Bayesian Approach

Giovanni Parmigiani
Johns Hopkins University, USA

JOHN WILEY & SONS, LTD

Copyright © 2002   John Wiley & Sons, Ltd
Baffins Lane, Chichester,
West Sussex, PO19 1UD, England

National   01243 779777
International (+44) 1243 779777

e-mail (for orders and customer service enquiries): cs-books@wiley.co.uk

Visit our Home Page on http://www.wiley.co.uk or http://www.wiley.com

Other Wiley Editorial Offices

John Wiley & Sons, Inc., 605 Third Avenue,
New York, NY 10158-0012, USA

WILEY-VCH Verlag GmbH
Pappelallee 3, D-69469 Weinheim, Germany

John Wiley & Sons Australia, Ltd
33 Park Road, Milton, Queensland 4064, Australia

John Wiley & Sons (Canada) Ltd, 22 Worcester Road
Rexdale, Ontario, M9W 1L1, Canada

John Wiley & Sons (Asia) Pte Ltd, 2 Clementi Loop #02-01,
Jin Xing Distripark, Singapore 129809

Library in Congress Cataloging-in-Publication Data

Parmigiani, G. (Giovani)
    Modeling in medical decision making: a Bayesian approach / G. Parmigiani.
        p.; cm.–(Statistics in practice)
    Includes bibliographical references and index.
    ISBN 0-471-98608-9 (alk. paper)
        1. Medicine – Decision making – Statistical methods. 2. Bayesian statistical decision theory. I. Title. II. Statistics in practice (Chichester, England)
        [DNLM: 1. Decision Supported Techniques. 2. Bayes Theorem. 3. Decision Making, Computer-Assisted. 4. Models, Statistical. W. 26.55.D2 P255mm 2002]
    R723.5. P37 2002
    610'.7'27 – dc21   2001045536

British Library Cataloguing in Publication Data

A catalogue record for this book is available from the British Library

ISBN 0-471-98608-9

Produced from LaTeX files supplied by the author, processed by Laserwords Private Limited, Chennai, India.
Printed and bound in Great Britain by TJ International, Padstow, Cornwall.
This book is printed on acid-free paper responsibly manufactured from sustainable forestry, in which at least two trees are planted for each one used for paper production.

Contents

Part II   Case Studies   121

Preface

Medical decision making is evolving toward an evidence- and model-based scientific discipline, as a result of the increasingly large and complex body of information that needs to be brought to bear in making decisions in health care. While decision makers are generally reluctant to delegate the decision process entirely to model-based formalism, they are increasingly appreciating the role of quantitative modeling, and relying on formal approaches to make more informed decisions.

In parallel, fast computing and recent progress in simulation-based inference have led to a host of new and powerful statistical tools, many of which have direct implications for medical decision making problems. In particular, simulation-based Bayesian methods are proving successful, as they are based on simple and universal notions, and can help address some of the most pressing practical and ethical concerns arising in medical decision problems. These include:

- combining information from diverse sources;
- incorporating expert judgment in a natural, nonbinding way;
- addressing quantitatively all relevant sources of uncertainty in the conclusions;
- unifying the thinking about study design, inference and decision making;
- incorporating new information as it accrues sequentially, to maximize the efficiency with which new knowledge is translated into clinical practice.

My goal in writing this book is to help bridge the gap between simulation-based Bayesian statistical methods and their use in medical decision making. My hope is that both statisticians and applied decision modelers may find it helpful. Statistical analysis can be improved with a clearer focus on the relevant decision problems; and decision analysis can be improved by using more effective statistical tools to assess values, outcomes and uncertainties, and by more directly integrating the process of information synthesis into decision modeling.

The book's potential readership includes both students and practitioners. A first-year doctoral or master's student in disciplines such as statistics, economics, business, health policy, epidemiology, and biostatistics should be able to follow the book throughout. Practitioners with some experience in decision analysis or statistical data analysis should be able to follow the general ideas, which are usually described in intuitive terms before being formalized. Most of the discussion only requires a basic statistical background, such as a calculus-based introductory class in statistics or biostatistics.

The book is divided into two parts, each consisting of three chapters. Part I provides a concise overview of the important principles and techniques, focusing on inference, formal decision making, and simulation. Part II gives a detailed account of three case studies, based on collaborative work of which I have been one of the authors. In every case, the original material has been expanded, with three goals in mind: to provide additional medical and/or policy background and give details of the process of building the model and identifying the appropriate statistical tools; to provide further background on the statistical techniques and solutions via simplified versions of the case study, and clarify the workings of individual techniques; and to provide a more detailed account of the implementation than is typically possible in a journal article.

Of the three case studies, two are the result of collaborative research. I also drew extensively on at least two additional major collaborative projects. The book would simply not exist without these collaborations. I have devoted much effort to revising the original journal articles that describe the case studies, to make them fit in the framework of the book, by modifying the notation, the graphical presentation and, in every instance, at least some aspects of the modeling and analysis. However, the substantive ideas that are presented, and sometimes the language describing them, emerged from collaborative work. I obviously am greatly indebted to my co-authors, not simply for what they indirectly contributed to this book, but also for teaching me most of the medical and statistical ideas that I am trying to convey.

More specifically, the genetic testing application discussed in Chapter 1 is based on BRCAPRO, a brainchild of the genius and vision of Don Berry. BRCAPRO profited from input from a number of participants in the Duke Specialized Program of Reasearch Excellence (SPORE) in breast cancer, led by Dirk Iglehart. Among these, I am especially grateful to Ed Iversen and Joellen Schildkraut. Thanks to the work of David Euhus, BRCAPRO has now hundreds of users. Many of their doubts, questions, objections, and suggestions have affected my way of presenting the material.

The illustrative stroke model of Chapter 3, called μSPPM, is a much simplified version of the Stroke Prevention Policy Model developed by the Stroke Patient Outcomes Research Team (PORT) at the Center for Clinical Health Policy at Duke. My discussion of it has been strongly influenced by the ideas of David Matchar and Greg Samsa, and also reflects contributions by Joe Lipscomb.

The migraine headache material of Chapter 4 is part of Francesca Dominici's PhD thesis and was originally developed in collaboration with Vic Hasselblad and Robert Wolpert, both of whom contributed substantially to the modeling presented in the chapter. I am also grateful to Doug McCrory for providing the data, which were abstracted from published randomized controlled trials as part of the literature review in support of Clinical Practice Guidelines funded by the Agency for Health Care Policy Research.

The axillary lymph node dissection decision analysis of Chapter 5 is the result of the collaboration of three statisticians (Don Berry, Claudia Tebaldi and myself), a surgical oncologist (Dirk Iglehart), a radiation oncologist (Len Prosnitz), and a clinical oncologist (Eric Winer). It too was part of the Duke SPORE in breast cancer. I am especially grateful to Len for patiently trying to teach me how to write in a way that may be understandable to clinicians. I hope that echoes of his inimitable prose made it into the book.

The breast cancer screening model of Chapter 6 grew out of my own dissertation, sparked by the late Morris DeGroot and nurtured by Jay Kadane. Morrie's lung cancer diagnosis gave me the initial impulse to direct my early work on optimization toward cancer prevention applications. Later, I collaborated on various aspects of screening modeling with Steve Skates and Marvin Zelen, whose ideas have influenced the discussion in the chapter. David Draper and Peter Freeman were discussants of a conference presentation of my breast screening model and gave very useful feedback.

Many people helped me write this book. Jay Kadane and Helen Ramsey made the book contract possible. Sharon Clutton and Rob Calver of Wiley provided much valued support. Shing Lee heroically read the entire book and gave me careful and insightful help. Heidi Ashih, Francesca Dominici, David Euhus, Steve Goodman, Lurdes Inoue, Harold Lehmann, and Hedibert Lopes read early versions of various chapters and gave very helpful comments. Many other people tolerated ursine or insolvent behavior throughout the process, but the list of those would really be too long to print!

Giovanni Parmigiani
Baltimore, USA

Part I

Methods

1

Inference

1.1 Summary

In this chapter we illustrate the fundamentals of Bayesian modeling with examples from medical decision making. A critical aspect of Bayesian modeling is that all relevant uncertainties are represented by probability distributions. In medical decision making there are different sources of uncertainty that need to be understood and quantified. These include variations in prognosis from patient to patient, sampling variability in experimentation, heterogeneity of results across different studies, imperfect expert knowledge of critical quantities or mechanisms, and so forth. There are of course differences among these uncertainties: these differences and their practical implications are recognized in Bayesian modeling. But mathematically, all uncertainties are handled using the same tool – probability.

The goal of this chapter is to help develop intuition for probabilistic inference, its meaning, its workings, and its practical implications for medical decision making. In keeping with the general approach of the book, which is to present statistical ideas and techniques as they arise from medical decision making situations, this chapter is based on the following four extended examples, or miniature case studies.

1. *Medical diagnosis.* Interpreting the result of a medical test in patient diagnosis is the prototypical example of probabilistic inference (Sox, 1986). We will consider the diagnosis of liver disease using a liver scan, based on data reported by Drum and Christacapoulos (1972). In this context, we will introduce the concepts of conditional probability, sensitivity, specificity, pre-test, and post-test probabilities. We will illustrate the role of these probabilities in quantifying the process of learning about a patient's disease status.

2. *Genetic counseling.* A critical component of genetic counseling is an assessment of the probability that the person being counseled carries a certain genotype, based on related phenotypes throughout his or her family tree (Murphy and Mutalik, 1969). Like medical diagnosis, this is an activity

that relies on probabilistic inference, but here the task is more complex, as it requires handling several unknowns, combining the contribution of several empirical observations, and incorporating external knowledge about the inheritance mechanism of the genotype of interest. We will study in detail a simplified version of the breast cancer genetic counseling model BRCAPRO (Parmigiani *et al.*, 1998).

3. *Sensitivity and specificity estimation.* The accuracy of a diagnostic test is often studied by performing the test on samples of patients of known disease status. We will reconsider the liver scan example in this light. Our focus will shift from making inferences about the disease or genotype status of a single individual to making inferences about the characteristics of a population, such as sensitivity and specificity. We will illustrate the fundamentals of statistical inference, including the concepts of statistical model, conditional independence and likelihood. We will make the case that conditional probability can be used to quantify uncertainty about population characteristics, and show how to account for this uncertainty in making predictions about patient status when the test characteristics are not known with certainty.

4. *Chronic disease modeling.* A useful starting point in understanding the efficacy of preventive interventions is modeling the natural course of the disease in the absence of such interventions. Our second example of inference will consider the parameters of a model that describes, in a simplified way, the history of patients at high risk of stroke (Parmigiani *et al.*, 1997). We will consider a hypothetical cohort and use this model as a tutorial example at several other points in Chapters 2 and 3.

While these examples are designed to introduce the important ideas, there is no attempt to be exhaustive. The material begins at a level that a reader with no prior knowledge of statistics should be comfortable with, but it is not intended to be a self – contained introduction to probability and it proceeds quickly toward relatively advanced techniques. Readers whose knowledge of probability and statistics is somewhat rusty will find it useful to have access to a copy of DeGroot (1986). There are also many good introductory textbooks on Bayesian inference, covering the basics in much greater depth. A classic, concise, and illuminating introduction is Edwards *et al.* (1963). A basic pre-calculus treatment, with many simple and insightful examples, is Berry (1996). Useful manuals at a technical level that is appropriate for developing the expertise necessary to carry out independently applications such as those discussed in later chapters include Lindley (1965), O'Hagan (1994), Gelman *et al.* (1995), and Carlin and Louis (2000). More advanced technical treatments are provided by Hartigan (1983), Bernardo and Smith (1994), and Schervish (1995). While all these references include, to varying degrees, a discussion of the philosophical underpinnings of Bayesian modeling, there is also an extensive literature concerned specifically with foundational matters, many of which are very relevant to the medical context. Useful entry points are Barnett (1982) and Goodman (1999a; 1999b). Examples of Bayesian modeling applied to the medical sciences are now common. Two excellent sets of case studies, largely complementary to the material discussed here, are Kadane (1996) and Berry and Stangl (1996). A skeptical view that challenges the uncritical acceptance of Bayes' rule in clinical decision making is provided by Feinleib (1977).

1.2 Medical diagnosis

1.2.1 Sensitivity, specificity and positive predictive value

Diagnostic tests are medical procedures used to discern healthy individuals from those who are ill. Some tests can do this very reliably, while most will make errors. For example, radiological procedures for the detection of cancer may miss small or hidden tumors, and can also raise concerns about masses that are not tumors at all. Probabilistic inference deals with drawing conclusions about an individual who has been tested based on the result of the test, on what is known about the accuracy of the test, and on what is known more broadly about the medical condition that the test is attempting to reveal. To make our discussion more concrete, we will consider the ability of hepatic scintigraphy (a liver scan) to detect liver disease. The concepts we introduce are quite general and apply to any binary diagnostic test.

Hepatic scintigraphy is a commonly used imaging procedure for detecting abnormalities in the liver. Drum and Christacapoulos (1972) report data on 344 patients who underwent scintigraphy and were later examined further by autopsy, biopsy, or surgical inspection, for direct determination of the liver pathology, independent of the liver scan results. Patients were referred to a liver scan primarily to search for metastatic tumor, but also for the evaluation of hematologic disorders, the evaluation of parenchymal liver disease, the search for liver abscesses, and other reasons. We use 'liver disease' to indicate any of these conditions. Table 1.1 summarizes the correlation between liver scan and pathologic findings. It is formed by classifying patients according to whether the liver scan is positive (T+) or negative (T−), and according to whether the disease is truly present (D+) or absent (D−). There are four possible combinations: each is represented by a cell in the table. For example, patients in the top left-hand cell (D+, T+) have liver disease and also tested positive, while patients in the bottom right-hand cell (D−, T−) do not have liver disease and tested negative.

For the remainder of this section, let us assume that the information in Table 1.1 is perfectly representative of the population of interest. Ideally, we would like a representative sample of patients who are referred to a liver scan. What we have is a sample of patients whose liver was further examined at some later point. The two populations may be similar with regard to liver test

Table 1.1 Two-way classification of liver scan (hepatic scintigraphy) results according to the outcome of the liver scan and the actual state of the patient's liver.

		Liver disease D+	No liver disease D−	Total
Abnormal liver scan	T+	231	32	263
Normal liver scan	T−	27	54	81
Total		258	86	344

performance, but are probably different with respect to the frequency of liver disease. Drum and Christacapoulos (1972) indicate that about 63% of the patients referred to a liver scan showed evidence of disease from either clinical or pathological findings, which is lower than the 75% showing pathological evidence in the sample of Table 1.1. However, for clarity of presentation, we will not consider this limitation further. Also, while 344 patients constitute a relatively large sample, uncertainty about test accuracy remains after the study. In Section 1.4, we will discuss quantifying the uncertainty arising from the fact that knowing about this sample of 344 patients is not exactly the same as knowing about the whole population.

The data in Table 1.1 can help a physician answer two important, related, but different questions: How accurate is a liver scan? And how should the result of a liver scan assist in the clinical management of a patient? We will consider these two questions in turn. Both require the use of conditional probabilities, although of different kinds.

Imagine randomly drawing a patient from a set like that of Table 1.1. The probability of that patient having both liver disease and a positive liver scan is equal to the number of patients in the top left-hand cell, divided by the size of the set, that is, $231/344 = 0.672$. In terms of the two events of interest here – the outcome of the test and the disease status – this will be denoted as $P(T+, D+)$, and termed *joint probability*, because it refers to the concomitant, or joint, occurrence of the two events. Table 1.2 introduces a notation for the joint probability of a patient falling into each of the four categories defined in Table 1.1. For example, a indicates $P(T+, D+)$. Table 1.2 also includes row and column totals, which can be interpreted as probabilities of events concerning the test alone and the disease alone, respectively. For example $a + b$ is the probability of an abnormal (positive) liver scan, irrespective of disease status. These probabilities are sometimes called *marginal*, since they are row or column totals from the table, which usually appear in the margins. In symbols, $a + c = P(T+) = 0.765$.

Table 1.2 Two-way table of probabilities of test and disease outcomes for the population of liver scan patients of Table 1.1. Entries are expressed as proportions of the population total, so each entry is equal to the corresponding entry of Table 1.1 divided by the total number of patients, 344.

		Liver disease D+	No liver disease D−	Total
Abnormal scan	T+	$a = 0.672$	$c = 0.093$	$a + c = 0.765$
Normal scan	T−	$b = 0.078$	$d = 0.157$	$b + d = 0.235$
Total		$a + b = 0.75$	$c + d = 0.25$	$a + b + c + d = 1$

It is standard to answer our first question, how accurate a liver scan is, by considering both *specificity*, or the ability of the test to recognize healthy patients, and *sensitivity*, or the ability of the test to recognize diseased patients. There are two types of correct classification in Table 1.2: classifying a diseased patient as diseased ((T+, D+) or *true positive*) and classifying a healthy patient as healthy ((T−, D−) or *true negative*). Quantitatively, sensitivity is defined as the *true positive rate*, or the probability of classifying a diseased individual as diseased. Specificity is the *true negative rate*, or the probability of classifying a healthy individual as healthy.

True negative and true positive rates are examples of *conditional probabilities*, that is, probabilities that refer to subgroups of the population. For example, to calculate the true positive rate we focus on the diseased subset of the population, and compute the fraction testing positive among those. In symbols, this is denoted by $P(T+|D+)$, and is read as *probability of* T+ *given* D+. In the notation of Table 1.2, the true negative rate is

$$\alpha = P(T-|D-) = \frac{d}{c+d} = \frac{0.157}{0.25} = 0.628.$$

Similarly, the true positive rate is

$$\beta = P(T+|D+) = \frac{a}{a+b} = \frac{0.672}{0.75} = 0.895.$$

Specificity and sensitivity are thus commonly indicated by α and β respectively, and we will follow this notation in the remainder of the chapter. Sensitivity and specificity are directly relevant in a range of important decisions, such as whether to recommend a diagnostic test, or whether to approve or purchase a diagnostic testing machine. Later we will explore how they contribute to clinical decisions as well.

Incidentally, the expressions above are examples of a general relationship between conditional, joint, and marginal probabilities. In the case of the events T+ and D+, this relationship takes the form

$$P(\text{T+}|\text{D+}) = \frac{P(\text{T+}, \text{D+})}{P(\text{D+})},$$

and it is usually considered to be the formal definition of conditional probability.

Let us now consider our second question, that is, how the result of a liver scan should assist in the clinical management of a patient. A natural way to translate this question in quantitative terms is to ask: What is the probability that the patient who tests positive is diseased? This is again a conditional probability, namely $P(\text{D+}|\text{T+})$, and is called the *positive predictive value*. There is an important difference between positive predictive value and sensitivity, which is $P(\text{T+}|\text{D+})$. Suppose that the liver scan is assessed immediately by a radiologist, so there are no delays. Sensitivity looks at a group of diseased patients who walk into the liver scan room, and answers the question: How many of them will have a positive scan? Positive predictive value looks at a group of patients who walk out of the liver scan room with a positive result, and answers the question: How many of those really have liver disease? The latter is the relevant quantity to assist clinical decision making about patients who test positive.

Because we know how to compute conditional probabilities from joint and marginal probabilities, we can easily determine the positive predictive value from Table 1.2. We need to focus on positive patients and compute fraction of diseased patients among those. Formally:

$$P(\text{D+}|\text{T+}) = \frac{a}{a+c} = \frac{0.672}{0.765} = 0.878.$$

In this calculation, the numerator is the same as that used in computing sensitivity, but the denominator is different. Because the two denominators are close in this example, sensitivity and positive predictive value are also close, but this need not be the case in general.

Similarly, we can look at the *negative predictive value*, or the probability of being healthy given a negative test, which is

$$P(\text{D--}|\text{T--}) = \frac{d}{b+d} = \frac{0.157}{0.235} = 0.667$$

In summary, quantifying the accuracy of a diagnostic test requires considering the two columns in Table 1.2 separately. Quantifying the implications of the test results requires considering the two rows in Table 1.2 separately. The row one considers depends on the outcome of the test.

Table 1.3 Two-way table of probabilities of test and disease outcomes for the population of liver scan patients of Table 1.1. The new notation simplifies the column totals but complicates the row totals.

		Disease D+	No Disease D−	Total
Positive test	T+	$\pi\beta$	$(1-\pi)(1-\alpha)$	$\pi\beta + (1-\pi)(1-\alpha)$
Negative test	T−	$\pi(1-\beta)$	$(1-\pi)\alpha$	$\pi(1-\beta) + (1-\pi)\alpha$
Total		π	$1-\pi$	1

1.2.2 Bayes' rule

We are now ready to consider the question of how sensitivity and specificity can contribute to clinical diagnosis, by examining how these quantities affect the positive predictive value. To simplify the notation, we will indicate by π the fraction of diseased individuals in the population, usually called *prevalence*. In the symbols of Table 1.2:

$$\pi = a + b.$$

The prevalence π is a property of the patient population.

Substituting π, α, and β for a, b, c, and d, we can rewrite Table 1.2 in terms of sensitivity, specificity, and prevalence, and obtain the expressions in Table 1.3. These let us represent positive and negative predictive values in terms of prevalence, sensitivity, and specificity. Let us indicate the positive predictive value by π^+. Then

$$\pi^+ = P(\text{D+}|\text{T+}) = \frac{a}{a+c} = \frac{\pi\beta}{\pi\beta + (1-\pi)(1-\alpha)}. \tag{1.1}$$

Using a similar argument, the fraction of diseased patients among patients testing negative is

$$\pi^- = P(\text{D+}|\text{T−}) = \frac{b}{b+d} = \frac{\pi(1-\beta)}{\pi(1-\beta) + (1-\pi)\alpha}, \tag{1.2}$$

so that the negative predictive value is $1 - \pi^-$. In the liver scan application $\pi^+ = 0.878$ and $\pi^- = 1 - 0.667 = 0.333$.

Both these expressions are examples of a general inference rule called Bayes' rule. They are just simple restatements of the probabilities in the table, but they can be used to answer a very important question: What can we say about a tested patient, if we know the accuracy of the test and the prevalence of the disease in the population from which the patient is drawn?

For example, consider testing volunteer blood donors for the HIV infection, using the enzyme immunoassay (EIA) test. In the late 1980s, this test had a sensitivity $\beta = 0.99$ and a specificity $\alpha = 0.999$. Yet, less than 10% of the positives would be subsequently confirmed to be HIV infections using a Western blot test (Barry *et al.*, 1986). This result is less surprising than it may appear, because the prevalence of HIV infection among volunteer blood donors was of the order of one in 10 000 ($\pi = 0.0001$). Using expression (1.1), we can determine the positive predictive value to be 0.09. This illustrates mathematically a fact which is well known to all good diagnosticians: the clinical significance of a positive result depends on the population on which it is applied. So sensitivity and specificity alone do not provide enough guidance to make clinical diagnoses: they cannot be converted into clinically relevant quantities without knowing something about disease prevalence.

Probabilities of disease represent the diagnostician's inference of the disease status of patients. The process of going from π to either π^+ or π^-, using Bayes' rule, quantifies how these inferences are updated in the light of the evidence provided by the liver scan. In this sense the transition from π to π^+ or π^- is taken to model the diagnostician's learning about the disease status of the patient. It is common terminology to call π an *a priori*, or *prior*, or *pre-test* probability, because that is the inference that would be made before, or in the absence of, the test result. Analogously, π^+ and π^- are called *a posteriori*, or *posterior*, or *post-test* probabilities, because they reflect the inference that would be made after, or in the light of, the test result.

The diagnostician's inference about the patient's liver status exemplifies a much broader approach to making probabilistic inferences about unknowns. In general, if we are interested in describing how knowledge about a hypothesis H is revised in the light of new evidence E, we can do so using Bayes' rule. The input to Bayes' rule are the prior probability of the hypothesis H and the probabilities of the evidence E conditional on the hypothesis being true and false, respectively. In this context, Bayes' rule takes the form

$$P(\text{H}|\text{E}) = \frac{P(\text{H})P(\text{E}|\text{H})}{P(\text{H} = \text{true})P(\text{E}|\text{H} = \text{true}) + P(\text{H} = \text{false})P(\text{E}|\text{H} = \text{false})}.$$
(1.3)

Expressions (1.1) and (1.2) are special cases in which we take H to be D+ and E to be T+ and T−, respectively.

Bayes' rule in medical diagnosis is discussed in may texts, including Lusted (1968) and Sox *et al.* (1988). More general versions of expression (1.3) are available for situations in which there are more than two possibilities for the hypothesis. For example, we could be interested in diagnosing several levels of liver disease, or we may be interested in discerning among several conditions that could give rise to the same symptoms. In that case the numerator of expression (1.3) is unchanged, but the denominator requires a sum with as many terms as there are possibilities. Discriminating among several conditions will not be explored here in detail, but it is quite interesting. A discussion aimed at a medical audience is provided by Gorry *et al.* (1978), who discuss how negative results can be evidence against one disease and evidence for another. A more general illustration in the context of Bayesian graphical models is given by Szolovits (1995).

Most of this chapter is devoted to discussing examples, extensions, and applications of probabilistic reasoning. In the remainder of this section, we will look at two more variations on the theme: odds and likelihood ratios, to investigate the effect of the sensitivity and specificity on our conclusions; and prior-to-posterior graphs, to investigate the effects of prevalence on our conclusions.

1.2.3   Odds, Bayes factors and the nomogram

Odds are a common way of expressing chances of events. The odds of an event are the ratio of the probability of the event happening to the probability of it not happening. For example, the odds of disease in the population of Table 1.1, or prior odds, are $\omega = \pi/(1-\pi) = 3$. Odds can be any positive number. Odds of 1, or 'even odds', correspond to a probability of 0.5. More generally, the probability π can be recovered from the odds by the relationship $\pi = \omega/(1+\omega)$.

What are the odds of disease given a positive test? With some simple algebra, we can rewrite π^+ as

$$\pi^+ = \frac{\frac{\pi}{1-\pi}\frac{\beta}{1-\alpha}}{1 + \frac{\pi}{1-\pi}\frac{\beta}{1-\alpha}}$$

and conclude that

$$\frac{P(D+|T+)}{P(D-|T+)} = \frac{\pi^+}{1-\pi^+} = \frac{\pi}{1-\pi}\frac{\beta}{1-\alpha}. \tag{1.4}$$

Using a similar argument,

$$\frac{\pi^-}{1-\pi^-} = \frac{\pi}{1-\pi}\frac{1-\beta}{\alpha}. \tag{1.5}$$

These expressions give Bayes' rule in a simple multiplicative form: the posterior odds are equal to the prior odds $\pi/(1-\pi)$ multiplied by the ratio $\beta/(1-\alpha)$, which is sometimes called the *weight of evidence* or *Bayes factor*. In this simple setting the Bayes factor coincides with what is known as the *likelihood ratio*.

In medical diagnosis using a test with binary outcome, the Bayes factor is the ratio of the probability of observing the test outcome actually observed if the patient is diseased, to the same probability if the patient is not diseased. For example, if a test has sensitivity 90% and specificity 55%, we are twice as likely to observe a positive outcome if the patient is diseased as we are if the patient is healthy. So in this case the weight of evidence provided by a positive test in favor of disease is 2. Generally, in view of expressions (1.4) and (1.5), the Bayes factor of a positive result (i.e. $\beta/(1 - \alpha)$) and the Bayes factor of a negative result (i.e. $(1 - \beta)/\alpha$) are intuitive ways of expressing sensitivity and specificity information. They measure directly the impact of empirical observation of the test on the predictive values. An accurate test will have a high value of $\beta/(1 - \alpha)$ and a low value of $(1 - \beta)/\alpha$.

In the liver scan application,

$$\frac{\pi^+}{1 - \pi^+} = \frac{\pi}{1 - \pi} \frac{\beta}{1 - \alpha} = 3 \left(\frac{231/258}{1 - 54/86} \right) = 3 \times 2.406,$$

$$\frac{\pi^-}{1 - \pi^-} = \frac{\pi}{1 - \pi} \frac{1 - \beta}{\alpha} = 3 \left(\frac{1 - 231/258}{54/86} \right) = 3 \times \frac{1}{6}$$

so that a positive test has a weight of evidence of 2.406, or about 12 to 5, in favor of disease, while a negative test has a weight of evidence of 1 to 6 in favor of disease. Because the latter is less than 1, a negative test lends more support to absence of disease than it does to presence of disease, as is to be expected. After a positive scan we obtain the posterior odds by multiplying the prior odds by a factor between 2 and 3, while after a negative scan we obtain the posterior odds by dividing the prior odds by a factor of 6. Evidence provided by a negative has a greater impact than evidence provided by a positive. The two are the same only when sensitivity and specificity are the same.

By taking logarithms of the multiplicative rule (1.4) we can make an interesting observation:

$$\log \text{ posterior odds} = \log \text{ prior odds} + \log \text{ Bayes factor.} \qquad (1.6)$$

For a fixed Bayes factor, an increase of 1 in the log prior odds translates directly into an increase of 1 in the log posterior odds, no matter what the Bayes factor is. Similarly, for a fixed prior, an increase of 1 in the log Bayes factor translates directly into an increase of 1 in the log posterior odds, no matter what the prior is.

We can use this fact to construct the so-called nomogram. This is a graphical representation of expression (1.6) that is sufficiently easy to use that one could evaluate Bayes' rule on the fly in an operating room. An example is given in Figure 1.1. The marks on the vertical scales correspond to the pre-test (or prior) probability, the Bayes factor, and the post-test (or posterior) probability. Because of relationship (1.6) the vertical midpoint between the log prior and the negative log posterior is half the log of the Bayes factor. So Figure 1.1 can be used to determine any one of the three terms in (1.6) given the other two. For example, by placing a ruler over a value of a prior and a Bayes factor, one can read the corresponding posterior. An early nomogram appeared in an article by Fagan (1975). Since then it has come into common

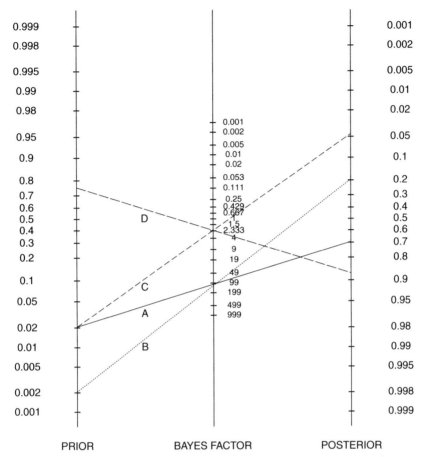

Figure 1.1   Nomogram for Bayes' rule. Line D represents the liver scan example. Lines A, B, C, and D correspond to points A, B, C, and D in Figure 1.2.

use in reporting information about diagnostic tests. With current computing technology, the nomogram has lost its appeal as 'the slide ruler of medical diagnosis', but it is still useful to play around with it to get a sense for the weight of evidence provided by the Bayes factor and for how it impacts different pre-test probabilities.

For example, in Figure 1.1, line D represents the liver scan application. The prior on the left scale is $\pi = 0.75$ and the Bayes factor on the middle scale is about 2.4. By joining those two points we can project the posterior to the value of 0.878. Contrast line D with line C, representing the result of applying a liver scan example to a population with a 2% prevalence: the positive predictive value would be barely 5%. The solid diagonal line A corresponds to a much more accurate test, characterized by a Bayes factor of 99 (e.g. $\beta = 0.99$ and

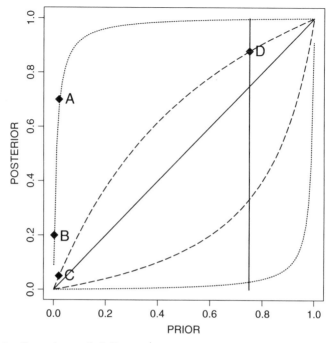

Figure 1.2   Posterior probabilities π^+ or π^- as a function of the prior probability π for two diagnostic tests. Lines A, B, C, and D in Figure 1.1 correspond to points A, B, C, and D here.

$1 - \alpha = 0.01$) and applied to a patient population with a prevalence of 2%. The resulting positive predictive value is 70%. Contrast this with the dotted line B, representing the same test applied to a population with only 2 cases per thousand. The positive predictive value is now only 20%.

An alternative way of graphically representing the contributions of different sources of information in Bayes' rule is the prior-to-posterior graph of Figure 1.2. This displays the posterior probabilities π^+ or π^- from expressions (1.1) and (1.2) as a function of the prior probability, for a fixed Bayes factor. A Bayes factor of 1, which corresponds to an uninformative test, will leave the prior unchanged. This case corresponds to the diagonal line in Figure 1.2. If a test is informative, it will be characterized by two lines, one above the diagonal, corresponding to a positive result, and one below the diagonal, corresponding to a negative result.

For example, in Figure 1.2, the dashed lines correspond to the liver scan application. The dashed line above the diagonal reflects evidence in favor of disease (positive test) and leads to post-test probabilities that are higher than the pre-test. The dashed line below the diagonal reflects evidence against disease (negative test) and leads to post-test probabilities that are lower than the pre-test. Using this graph we can derive the clinical significance of positive and negative liver scans in other populations. For example, Drum and Christacapoulos (1972) indicate that about 63% of the patients who underwent a liver scan showed evidence of disease from either clinical or pathological findings. By setting the prior on the horizontal scale to 0.63 we can adjust the positive predictive value accordingly. Also, if we know the specific cause that led to the indication of a liver scan (search for liver metastases versus search for parenchymal liver disease, etc.), we may be able to use prior probabilities that are specific to such subgroups.

Figure 1.2 also illustrates the effect of test accuracy on predictive values. The better the test, the further away the predictive value lines will be from the diagonal. For example, increasing the Bayes factors to 99 (dotted line) and 1/99 will push the graphs further away from the diagonal. Consider the case of a prior probability of 0.75, as in the liver scan application. All possible posterior probabilities are represented by the vertical line in Figure 1.2. The liver scan results will revise the prior to either $\pi^+ = 0.878$ or $\pi^- = 0.333$, while the hypothetical test represented by the dotted line would revise the prior more markedly.

1.3 Genetic counseling

Our next example of probabilistic inference arises in the counseling of individuals who are concerned about the possibility of being genetically predisposed to inheritable diseases, such as breast and colon cancer or Alzheimer's disease. After some discussion of the medical background, we examine in detail a model for genetic counseling of women with a family history of breast cancer, and use it to illustrate how one can extend the simple diagnostic paradigm described in the previous section to situations in which we have several unknowns and several related empirical observations that provide evidence about the diagnosis to be made. Our discussion is based on the BRCAPRO model (Berry *et al.*, 1997; Berry and Parmigiani, 1997; Parmigiani *et al.*, 1998).

1.3.1 Genetic susceptibility to breast and ovarian cancer

Both breast and ovarian cancer cluster in families. In the 1980s researchers identified the hereditary nature of some breast and ovarian cancer clusters (Lynch *et al.*, 1984), and suggested the presence of autosomal dominant susceptibility genes (Claus *et al.*, 1990). In the 1990s our understanding of inherited susceptibility to breast cancer made further progress, with the identifications of two of these genes: *BRCA1* and *BRCA2*. Approximately 2% of breast cancers and 10% of ovarian cancers are believed to occur in women who carry an inherited deleterious mutation of *BRCA1*, on chromosome 17q (Miki *et al.*, 1994; Futreal *et al.*, 1994). A smaller proportion of breast cancer is attributable to inherited deleterious mutations of *BRCA2*, on chromosome 13q (Wooster *et al.*, 1995). Mutations are rare, occurring in less than 0.2%

of women (Ford and Easton, 1995). But accruing evidence is confirming that women with mutations are very likely to develop breast or ovarian cancer or both, and to do so at relatively young ages (Easton *et al.*, 1995, Narod *et al.*, 1995, Struewing *et al.*, 1997).

The isolation of the *BRCA1* and *BRCA2* genes allows direct testing of individuals for the presence of a mutation (Weber, 1996). For women who have a family history of cancer but are not affected themselves, genetic testing may provide important information about cancer risk, and the risk of transmitting a high susceptibility to their offspring (Warmuth *et al.*, 1997). Testing may also have implications for the future health of women with cancer. For example, women with breast cancer may be concerned about the risk of contralateral breast cancer and/or ovarian cancer.

Whether or not to be tested for *BRCA1* and *BRCA2* mutations is a complex decision. It depends on the woman's chance of carrying a mutation, as well as the effectiveness and cost of the testing procedure, the available prophylactic interventions, the effectiveness and negative effects of these interventions, the impact of testing on other family members, and the impact on the woman's ability to obtain insurance coverage (Meissen *et al.*, 1991; Schwartz *et al.*, 1995) and employment (Billings *et al.*, 1992). A positive test or simply the perception of a high risk can lead to aggressive management, ranging from more frequent mammograms to bilateral mastectomy, again with substantive consequences on a woman's life. In this scenario, there is great interest in providing useful information that can help women decide if they indeed want to undergo testing.

Recognizing these problems, the National Action Plan on Breast Cancer and American Society of Clinical Oncology (1997) have developed informative material for the education of physicians on issues of genetic testing. They stress that

> the decision whether to undergo predisposition genetic testing hinges on the adequacy of the information provided to the individual in regards to the risks and benefits of testing, possible testing outcomes, sensitivity and specificity of the test being performed, cancer risk management strategies, and the right of choice. Individuals should be provided with the necessary information to make an informed decision in a manner that they can understand.

A critical step in counseling a woman facing these decisions is an accurate evaluation of the probability that she carries a mutation.

1.3.2   Individualized probabilities

While there is evidence that other inheritable factors may affect breast cancer, *BRCA1* and *BRCA2* are associated with a large fraction of the cases that are attributable to genetic susceptibility (Newman *et al.*, 1997; Weber, 1998). This means that despite the rarity of mutations, they are likely to be present in families that have multiple occurrences of breast or ovarian cancer. A family history of these diseases is a strong indicator of whether a mutation is present in the family, and in a particular family member.

In fact, the chance of carrying a genetic mutation varies markedly from woman to woman, depending on the family history of breast and related cancers. Many aspects of a woman's family history are important in determining her chance of being a carrier, such as the exact relationships of all family members, including both affected and unaffected members, the ages at diagnosis of the affected members, and the current ages of the unaffected members. Particular relationships of family members with cancer and also without cancer can have a substantial impact on the probability of carrying a susceptibility gene. Ages at diagnosis of affected family members and their types of cancer are also important. An affected woman with several cancers in her family can have a probability of carrying a mutation that ranges from less than 5% to close to 100% (Berry *et al.*, 1997). A woman with two primary cancers can have a probability of carrying a mutation in excess of 80%, even with no other information about family history.

This complexity makes the provision of accurate carrier probability information something of a challenge. In response to this challenge, several quantitative models for determining carrier probabilities have been proposed. Model-based prediction is emerging as an efficient, potentially effective process that can enable the delivery of accurate individualized information about genetic testing to a larger audience. Quantitative models are currently being used in a range of counseling and clinical activities. Materials distributed to women that are considering genetic testing often include model-based predictions (Myriad Genetics, 2001; Bluman *et al.*, 1999).

In addition to clinical and counseling applications, model-based approaches are being used in a variety of scientific investigations (Blackwood and Weber, 1998). Their roles include: to help in the study of characteristics of *BRCA1* and *BRCA2* carriers in populations in which genetic testing has not been carried out, or has only partially been carried out (Schildkraut *et al.*, 1997; Iversen *et al.*, 1997; Smith *et al.*, 1996; Hartmann *et al.*, 1999); to guide the selection of high-risk families for the study of risk, and for the assessment of prevention strategies; and the study of the quality of care and counseling of high-risk women (Iglehart *et al.*, 1998).

Broadly speaking, two general modeling approaches have been used so far; here we will label them 'empirical' and 'Mendelian'. Empirical approaches build models using statistical modeling techniques directly on pedigree data for tested individuals: the test results constitute the response variable(s); pedigree features provide the predictor variables. Candidate features are often extracted from pedigrees based on clinical and epidemiological expertise. Highly predictive features are then identified using statistical variable selection techniques. Examples include Shattuck–Eidens *et al.* (1997), Couch *et al.* (1997), Frank *et al.* (1998), and Hartge *et al.* (1999).

By contrast, Mendelian models use background knowledge about the autosomal dominant inheritance of the genes to formulate explicit genetic models for predicting carrier status. This is done in two steps: estimation of 'genetic parameters' comprising the mutations' population prevalences and penetrance functions; and application of Bayesian prediction to transform the genetic parameters into carrier probabilities (Murphy and Mutalik, 1969; Elston and Stewart, 1971; Szolovits and Pauker, 1992; Offit and Brown, 1994). Penetrances and prevalences may be estimated directly or abstracted from published work.

Here we discuss in detail how to build a basic Mendelian model considering just a single gene, to ease the task of illustrating the fundamental concepts. Our discussion is based on a simplified version of BRCAPRO (Berry *et al.*, 1997; Parmigiani *et al.*, 1998), which is a Mendelian model and computer program for finding the probability that an individual carries a germline mutation at *BRCA1* or *BRCA2*, based on his or her family's history of breast and ovarian cancer. In the current implementation of the model, the family history includes the counselee and her first- and second-degree relatives. (First-degree relatives are parents, siblings and offspring. Second-degree relatives are first-degree relatives of first-degree relatives.) For each member – including the counselee – the model processes information about whether the member has been diagnosed with breast cancer and, if so, age at diagnosis or, if cancer-free, current age or age at death. Similar data are processed for ovarian cancer if the family member is female. Here, we focus on simple family structures.

Our discussion comprises estimates of genetic parameters, the subject of Section 1.3.3; structural assumptions, such as inheritance mechanism and conditional independence, presented in Section 1.3.4; computing principles and algorithms, also presented in Section 1.3.4; and sequential use of Bayes' rule to incorporate both family history and the results of genetic tests in counseling, addressed in Section 1.3.7.

1.3.3 Genetic parameters

For simplicity, we consider a single hypothetical gene, which we call BRCA. We assume that there are only two types (or alleles) of BRCA genes: the normal type and the type conferring a genetic susceptibility, also called deleterious. Each individual carries two copies of the BRCA gene, each of which is inherited from one parent, by drawing at random from each parent's two copies. So each individual could carry 0, 1, or 2 deleterious alleles. As is the case with the real breast cancer susceptibility genes, we assume that BRCA is an autosomal gene. More specifically, we assume that carrying 1 or 2 deleterious alleles confers the same risk. This assumption has not yet been validated by empirical evidence because the BRCA genes are too rare for an extensive study of individuals with two deleterious copies. In view of this rarity this assumption is also not likely to influence the results of the prediction model.

We will use the notation g to indicate the BRCA genotype, and g_i to indicate the BRCA genotype of individual i. We will restrict attention to two possibilities: $g_i = 1$ if individual i has one or two deleterious alleles (carrier), and $g_i = 0$ if individual i has two normal copies (noncarrier).

The fraction of female carriers of BRCA genetic susceptibility mutations who develop breast and ovarian cancer by a given age is often called penetrance. We use the letter x for age of onset of breast cancer and y for the age of onset of ovarian cancer. So the penetrance curves are defined as:

$$B_0(x) = P\{\text{Breast cancer by age } x \mid g = 0\},$$
$$B_1(x) = P\{\text{Breast cancer by age } x \mid g = 1\},$$
$$C_0(y) = P\{\text{Ovarian cancer by age } y \mid g = 0\},$$
$$C_1(y) = P\{\text{Ovarian cancer by age } y \mid g = 1\}.$$

Similarly, considering yearly increments of the curves above:

$$b_0(x) = P\{\text{Breast cancer at age } x \mid g = 0\},$$
$$b_1(x) = P\{\text{Breast cancer at age } x \mid g = 1\},$$
$$c_0(y) = P\{\text{Ovarian cancer at age } y \mid g = 0\},$$
$$c_1(y) = P\{\text{Ovarian cancer at age } y \mid g = 1\}.$$

Here, we take the penetrance functions of the hypothetical BRCA gene to be a weighted average of the penetrance functions for *BRCA1* and *BRCA2* in the current implementation of BRCAPRO. These are derived by combining the estimates reported by Ford *et al.* (1998) and Struewing *et al.* (1997). Details are discussed by Iversen *et al.* (2000). The resulting curves are shown in Figure 1.3. For example, B_1 is the dashed curve in the top left-hand panel, while b_1 is the dashed curve in the bottom left-hand panel. These functions will help us construct a likelihood function (i.e. the probability of the empirical evidence given the unknown genotype) for making inference on the genotype.

The other component that is needed for Bayes' rule is the prevalence of carriers of deleterious alleles in the population, $\pi = P\{g = 1\}$. In our example we hypothesize that the women being counseled are Ashkenazi Jews, an ethnic group characterized by an especially high prevalence of deleterious alleles. We derive estimates of mutation prevalence among the Ashkenazim from Oddoux *et al.* (1998) and Roa *et al.* (1996), both of which are population-based studies. The combined allele frequency for deleterious mutations of *BRCA1* and *BRCA2* is $\phi = 0.012\,89$. The probability of carrying at least one copy of a deleterious allele is then $\pi = \phi^2 + 2\phi(1 - \phi) = 0.025\,62$. The resulting prior odds of being a carrier are $\pi/(1 - \pi) = 0.0263$.

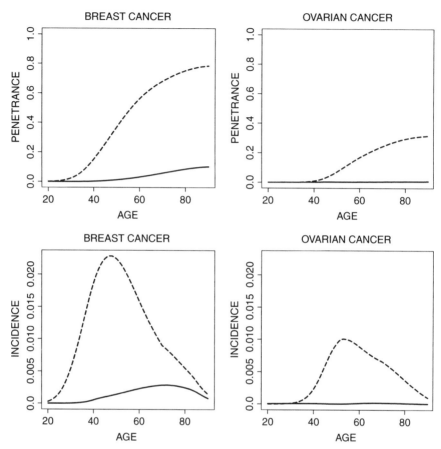

Figure 1.3 Hypothetical penetrance curves for the BRCA gene. The top row portrays the fraction of women developing breast and ovarian cancer by a given age, for carriers of deleterious BRCA mutations (dashed) and noncarriers (solid). The bottom row expresses the same information in terms of the number of cases per year at a given age. The curves at the bottom of each graph are obtained by taking yearly increments of the curves at the top.

1.3.4 Bayes' rule in genetic counseling

Our first example is that of a woman, say Anne, seeking counseling because of her mother's history of breast cancer. The family tree is pedigree A of Figure 1.4. Anne is the individual at the bottom of the pedigree. We will use $i = 1$ for Anne and $i = 2$ for her mother. Our goal is to compute the probability that Anne carries a deleterious BRCA allele. We will build our carrier probability calculation from simple to more complex. Initially we ignore ovarian cancer, and suppose that deleterious BRCA alleles only affect breast cancer risk.

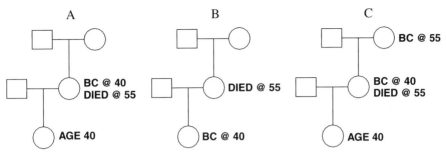

Figure 1.4   As customary in pedigree graphs, circles indicate females and squares indicate males. Offspring are below their ancestors. Horizontal links connect mates and vertical links represent parent–child relations. The graph also shows the complete cancer history and age information of female members of the family. Breast cancer cases are indicated by BC and ovarian cancer cases by OC.

To set the stage, we begin by deriving the probability that Anne is a BRCA carrier based on her own history only. We have two facts to consider: first, that Anne is an Ashkenazi, and we know the prevalence of deleterious BRCA alleles among Ashkenazim; and second, that Anne lived to be 40 without a diagnosis of breast cancer – in our notation $x_1 > 40$. The quantity of interest is

$P\{$Anne has deleterious BRCA allele | Anne has no breast cancer at 40$\}$

or, in the more formal notation,

$$\pi(g_1 = 1 \mid x_1 > 40).$$

To compute this probability we can use the information about ethnicity to specify a prior probability $\pi(g_1)$, and the information about family history to construct a Bayes factor. This situation mirrors that of Section 1.2: the presence of a deleterious allele is playing the same role as the presence of liver disease, and the personal history is playing the same role as the outcome of the liver scan. So, by analogy with expression (1.4), the posterior odds of carrying a deleterious allele are

$$\frac{\pi(g_1 = 1 \mid x_1 > 40)}{\pi(g_1 = 0 \mid x_1 > 40)} = \frac{\pi}{1 - \pi} \frac{p(x_1 > 40 \mid g_1 = 1)}{p(x_1 > 40 \mid g_1 = 0)}, \tag{1.7}$$

where the right-hand side of the equation is the product of the prior odds and the Bayes factor. Both the numerator and denominator of the Bayes factor can be determined directly from the information in Figure 1.3. Specifically, the top left-hand panel provides the probabilities $B_1(40)$ and $B_0(40)$ of developing breast cancer by age 40 for carriers and noncarriers, respectively. So the Bayes factor is

$$\frac{p(x_1 > 40 \mid g_1 = 1)}{p(x_1 > 40 \mid g_1 = 0)} = \frac{1 - B_1(40)}{1 - B_0(40)} = \frac{1 - 0.1509}{1 - 0.002\,318} \approx 0.85.$$

As expected, this provides weak, but not completely negligible, evidence against the possibility that the woman is a carrier. The prior odds decrease from 0.026 to 0.022 as a result of knowing that Anne was not diagnosed before age 40.

If instead Anne had been diagnosed at age 40 with breast cancer, we would need to consider the probabilities of diagnosis $b_1(40)$ and $b_0(40)$ for carriers and noncarriers, displayed in the bottom left-hand panel of Figure 1.3. The Bayes factor would be

$$\frac{p(x_1 = 40 \mid g_1 = 1)}{p(x_1 = 40 \mid g_1 = 0)} = \frac{b_1(40)}{b_0(40)} \approx 34.96,$$

leading to close to even odds that Anne is a carrier.

We now consider how to incorporate the information that Anne's mother was diagnosed with breast cancer at age 40, that is, $x_2 = 40$. The quantity of interest is

$$\Pr\{\text{Anne has deleterious BRCA allele} \mid \text{family history}\}$$

or, in the more formal notation,

$$\pi(g_1 = 1 \mid x_1 > 40, x_2 = 40).$$

Similarly to (1.7), the posterior odds of carrying a deleterious allele are

$$\frac{\pi(g_1 = 1 \mid x_1 > 40, x_2 = 40)}{\pi(g_1 = 0 \mid x_1 > 40, x_2 = 40)} = \frac{\pi}{1 - \pi} \frac{p(x_1 > 40, x_2 = 40 \mid g_1 = 1)}{p(x_1 > 40, x_2 = 40 \mid g_1 = 0)}. \tag{1.8}$$

Here the Bayes factor incorporates evidence from the whole family history, and is no longer so simple to evaluate: it involves two events rather than one, and the events are likely to be connected. Before we can make progress we need to make an important additional modeling assumption. This in turn requires a digression about the statistical concepts of independence and conditional independence.

1.3.5 Conditional independence

Two events are called statistically independent if knowledge of one does not change the probability of the other. For example, in the situation of Section 1.2, the outcome of a liver test may not depend on the presence of, say, hypertension. If the probability of a positive liver test in the overall population is the same as the probability of a positive liver test in the subpopulation of patients with hypertension, then knowing about a patient's hypertension does not affect the probability of a positive liver test, and the two events are said to be independent.

If we use H for hypertension, and D for liver disease, independence translates into

$$P(D|H) = P(D).$$

This definition appears asymmetric at first glance. Consider $P(H|D)$. By substituting $P(D, H)/P(H)$ for $P(D|H)$, we can also write independence as

$$P(D, H) = P(D)P(H) \qquad (1.9)$$

or

$$P(H|D) = P(H),$$

illustrating that independence is symmetric.

It is not uncommon for events to be independent in subgroups of a population, but not in the population at large. For a simple fictional example, consider a population consisting of two ethnic groups. Imagine that within each of the groups hypertension and hair loss are independent. This means that within an ethnic group, the fraction of hypertensive patients is the same among patients with and without hair loss. This illustrates one of the most important concepts in statistical modeling: that of conditional independence. Hypertension and hair loss are independent when we condition on ethnicity. If L is hair loss and E ethnicity, conditional independence is formally expressed as

$$P(H, L|E) = P(H|E)P(L|E),$$

which is simply expression (1.9) applied to a specific subgroup.

Now imagine that one of the ethnic groups has a much higher prevalence of both hypertension and hair loss. Then, in the population at large, the fraction of hypertensive patients is larger among patients with hair loss. The different ethnicities in the population and the different prevalences determine a dependence between hypertension and hair loss in the population at large, or more technically in the marginal distribution of the two features. This happens because hair loss provides information about ethnicity and thus, indirectly, about hypertension.

1.3.6   Bayes factors for family history

Returning to the genetic counseling example, we will use a specific conditional independence assumption to simplify the calculation of the Bayes factor of family history in expression (1.8). Cancer histories of mothers and daughters are clearly not independent in the population of all Ashkenazi families: we know that breast cancer tends to cluster in families. Is this clustering entirely due to inheritable susceptibility via the BRCA gene, or are there other factors that determine clustering? One way of thinking about this question in terms of conditional independence is to look at the subpopulation of women with known deleterious alleles and ask: Is there a relationship between cancer in those women and cancer in their mothers? In what follows we assume that the answer is no. Formally, we assume that the cancer histories of family members are conditionally independent given the genotype. Anne's mother's history of cancer informs us about Anne's risk only by informing us about her own genotype, which in turn affects Anne's genotype, and thus Anne's cancer risk.

This assumption of conditional independence of phenotypes (cancer histories) given genotype at a single locus is unlikely to be literally true. There may be other genes, or shared environmental factors, such as exposure of an entire family to radiation or carcinogenic pesticides, that determine familial clustering of cancer cases. However, if the BRCA gene is responsible for the majority of such clustering, the assumption of conditional independence may be reasonable. As in all modeling exercises, no model literally represents reality: some simplification is necessary to make progress, and some is useful in revealing important structure. How much simplification is too much is to be judged based on the accuracy of the model's results, and the costs of making the model more complex.

Our conditional independence relationships are illustrated in Figure 1.5. Circles represent the random variables in question. Arrows represent dependences among the variables. The directions of arrows represent natural ways of specifying conditional probabilities. For example, it is simple to specify conditional probabilities of the genotype of the daughter given the mother's, and to specify the probability of cancer-related events given the genotype.

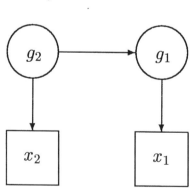

Figure 1.5 Graph of the conditional independence relationship among random variables in the genetic counseling problem, in pedigrees A and B of Figure 1.4. Quantities in rectangles are observed, while quantities in circles are unobserved.

The directions of arrows could also be attributed a causal connotation (which would be plausible in this specific example), but that does not have to be the case. Whatever the direction, arcs can be thought of as conduits of information. Our conditional independence assumption is reflected in the lack of a direct link between x_1 and x_2.

We are now ready to formalize the conditional independence conditions as:

$$p(x_1, x_2|g_1, g_2) = p(x_1|g_1, g_2)p(x_2|g_1, g_2) = p(x_1|g_1)p(x_2|g_2). \quad (1.10)$$

The first equality expresses the conditional independence of the individuals' phenotypes given the whole set of genotypes in the pedigree. It is analogous to the hypertension/hair loss example, with the two cancer histories replacing hypertension and hair loss, and the pair of genotypes replacing the ethnic group. The second equality reflects the additional assumption that an individual's phenotype depends on the relative's genotypes only via his or her own, as illustrated in Figure 1.5.

Using this factorization we can make progress toward evaluating the Bayes factor for family history. The two terms of the right-hand side of (1.10) can be evaluated directly by using the information in Figure 1.3, as we did in the simpler case of Section 1.3.4. For example, for pedigree A, we have

$$
\begin{aligned}
p(x_1 > 40, x_2 = 40 \mid g_1 = 1, g_2 = 1) \\
= p(x_1 > 40 \mid g_1 = 1)p(x_2 = 40 \mid g_2 = 1) \\
= [1 - B_1(40)]b_1(40) = 0.849 \times 0.0187 = 0.0159 \quad (1.11)
\end{aligned}
$$

and

$$
\begin{aligned}
p(x_1 > 40, x_2 = 40 \mid g_1 = 1, g_2 = 0) \\
= p(x_1 > 40 \mid g_1 = 1)p(x_2 = 40 \mid g_2 = 0) \\
= [1 - B_1(40)]b_0(40) = 0.849 \times 0.000\,534 = 0.000\,453, \quad (1.12)
\end{aligned}
$$

and so forth.

Our last obstacle is that we cannot work with $p(x_1, x_2|g_1, g_2)$ directly because in computing the numerator and denominator of the Bayes factor we cannot condition on the mother's genotype. Instead we need $p(x_1, x_2|g_1)$. In other words, we have here two unknowns, but only one is the focus of investigation. An argument similar to the one we have used to derive the denominator of Bayes rule in Section 1.2 comes in handy here. The idea is to derive $p(x_1, x_2|g_1)$ by a weighted average of $p(x_1, x_2|g_1, g_2)$'s. Consider the numerator of the Bayes factor first. Formally,

$$
\begin{aligned}
(x_1, x_2|g_1) = p(x_1, x_2, g_2 = 0|g_1) + p(x_1, x_2, g_2 = 1|g_1) \\
= p(x_1, x_2|g_2 = 0, g_1)p(g_2 = 0|g_1) \\
+ p(x_1, x_2|g_2 = 1, g_1)p(g_2 = 1|g_1), \quad (1.13)
\end{aligned}
$$

where the first equality is a marginalization over the g_2 dimension, and the second is an application of the definition of conditional probability.

This strategy is effective so long as we can determine the probabilities of the genotype of Anne's mother given Anne's own. Assuming Mendelian inheritance, we can determine these exactly. As we are studying the numerator of the Bayes factor, Let us assume that Anne is a carrier. What is the probability that her mother is as well? If we ignore the possibility that individuals in the pedigree carry two deleterious BRCA mutations, Anne can only have one deleterious allele, and this has the same probability of coming from her father or from her mother. So the probability that Anne's mother is a carrier is approximately one-half. On the other hand, if Anne is not a carrier (as we assume when evaluating the denominator of the Bayes factor), then her mother is very unlikely to be, but could be and have passed a normal copy of the gene. The probability of that happening is the probability of her 'other' allele being deleterious, that is, the allele frequency ϕ of deleterious mutations.

In reality the possibility of carrying more than one deleterious allele is not so remote among Ashkenazi families, especially when there are cancer cases on both sides of the family, but the bookkeeping of the Mendelian probabilities would lead us too far astray. BRCAPRO incorporates the correct probabilities, and you can refer to Parmigiani et al. (1998) or Weiss (1993) to see how that works.

In any event, using (1.13) and expressions (1.11) and (1.12), we obtain that the numerator of the Bayes factor is approximately

$$p(x_1 > 40, x_2 = 40 \mid g_1 = 1) = [1 - B_1(40)] \left\{ \frac{1}{2} b_1(40) + \frac{1}{2} b_0(40) \right\},$$

while the denominator is approximately

$$p(x_1 > 40, x_2 = 40 \mid g_1 = 0) = [1 - B_0(40)]\{\phi b_1(40) + (1 - \phi) b_0(40)\}.$$

The differences between the numerator and the denominator are in the conditioning population for the daughter (as before) and in the weights used to mix the two possible conditioning populations for the mother.

This exemplifies a very general feature of probabilistic inference, that is, the possibility of specifying models in terms of as many unknowns as are natural from a medical or scientific standpoint, and then getting rid of the unknowns that are not the focus of investigation by marginalization. The same approach could have been used to make inferences about the genotype of Anne's mother.

Evaluating the expressions above leads to the results of the first column of Table 1.4. The Bayes factor is about 10. Compare this to the value of about 35 that we obtained in Section 1.3.4 assuming that Anne herself had been diagnosed at age 40. Here the impact of Anne's mother's diagnosis on her own genotype is the same as that of Anne's diagnosis on her own genotype in

Section 1.3.4. But the uncertainty about whether Anne's mother has passed the gene to Anne, and Anne's health, all contribute to moderating the ultimate effect of this on the carrier probability.

Table 1.4 also considers pedigree A under a more elaborate model that incorporates ovarian cancer. Because Anne has not been diagnosed with ovarian cancer, and her mother lived to be 55 without a diagnosis of ovarian cancer, the relevant Bayes factor is

$$\frac{p(x_1 > 40, y_1 > 40, x_2 = 40, y_2 > 55 \mid g_1 = 1)}{p(x_1 > 40, y_1 > 40, x_2 = 40, y_2 > 55 \mid g_1 = 0)} =$$

$$\frac{[1 - B_1(40)][1 - C_1(40)] \left\{ \frac{1}{2}b_1(40)[1 - C_1(55)] + \frac{1}{2}b_0(40)[1 - C_0(55)] \right\}}{[1 - B_0(40)][1 - C_0(40)] \left\{ \phi b_1(40)[1 - C_1(55)] + (1 - \phi)b_0(40)[1 - C_0(55)] \right\}}.$$

From Table 1.4, we see that the Bayes factor is slightly reduced as a result of the lack of ovarian cancer diagnoses. The expression above relies on one additional conditional independence assumption: that, conditional on genotype, the times to diagnosis of breast and ovarian cancer are independent. This assumption is supported in the literature (Ford et al., 1998), although further investigation of the function of the BRCA genes may reveal violations.

Table 1.4 allows us to contrast pedigree A with pedigree B, in which the diagnoses are reversed. As expected, the impact on Anne's carrier probability of a diagnosis at age 40 is much greater (the Bayes factor is over 32) than the impact of the same diagnosis in her mother. However, the value of 32 is less than the value of about 35 that we obtained in Section 1.3.4 without incorporating information about the mother, who is healthy until age 55 in pedigree B. Again incorporating the information that neither of the two is diagnosed with ovarian cancer reduces the Bayes factor.

Our last calculation is for pedigree C, which is the same as pedigree A with the addition of a breast cancer diagnosis at 55 in Anne's maternal grandmother. Anne's grandmother will be individual 3. The conditional independence assumptions are now those of Figure 1.6, from which it is clear that the model has a recursive structure reflecting inheritance. So the same arguments that were used to relate Anne's mother's phenotype to Anne's genotype could be rephrased to relate Anne's grandmother's phenotype to

Table 1.4 Posterior probabilities, posterior odds, and Bayes factors for the pedigrees of Figure 1.4.

	Breast cancer only			Breast and ovarian cancer	
	A	B	C	A	B
Posterior probability	0.219	0.460	0.372	0.203	0.415
Posterior odds	0.280	0.852	0.591	0.255	0.710
Bayes factor	10.64	32.414	22.477	9.678	27.014

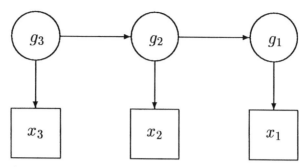

Figure 1.6 Graph of the conditional independence relationship among random variables in the genetic counseling problem, in pedigree C of Figure 1.4.

Anne's mother's genotype. Mathematically, we need to expand equation (1.13) recursively, by incorporating g_3 in the $p(x_1, x_2, x_3 | g_2, g_1)$ terms in the same way as we incorporated g_2 in the $p(x_1, x_2 | g_1)$ term. For example, we can write

$$p(x_1, x_2, x_3 | g_2 = 1, g_1)$$
$$= p(x_1, x_2, x_3 | g_3 = 0, g_2 = 1, g_1)p(g_3 = 0 | g_2 = 1)$$
$$+ p(x_1, x_2, x_3 | g_3 = 1, g_2 = 1, g_1)p(g_3 = 1 | g_2 = 1).$$

Without going into further detail, we give the expressions for the numerator and denominator of the Bayes factor, which are respectively:

$$p(x_1 > 40, x_2 = 40, x_3 = 55 \mid g_1 = 1)$$
$$= [1 - B_1(40)] \left\{ \frac{1}{2}b_1(40) \left\{ \frac{1}{2}b_1(55) + \frac{1}{2}b_0(55) \right\} \right.$$
$$\left. + \frac{1}{2}b_0(40) \left\{ \phi b_1(55) + (1 - \phi)b_0(55) \right\} \right\}$$

and

$$p(x_1 > 40, x_2 = 40, x_3 = 55 \mid g_1 = 0)$$
$$= [1 - B_0(40)] \left\{ \phi b_1(40) \left\{ \frac{1}{2}b_1(55) + \frac{1}{2}b_0(55) \right\} \right.$$
$$\left. + (1 - \phi)b_0(40) \left\{ \phi b_1(55) + (1 - \phi)b_0(55) \right\} \right\}.$$

Again, the results are given in Table 1.4. The grandmother's diagnosis lends substantial support to the presence of a genetic susceptibility in the family, and results in a Bayes factor which is over 22, more than doubling that of pedigree A. Note, however, that the addition of the mother's diagnosis has multiplied the original Bayes factor (0.85) by over 10. This difference is the result of both the closer proximity of the mother and the earlier age at diagnosis.

In summary, using Bayes rule, clinicians and counselors can integrate and interpret pedigrees, historical data, physical findings and laboratory data, providing individualized probabilities of various outcomes (Pauker and Pauker, 1987). The interpretation of these individualized probabilities is that they are approximately the proportion of Ashkenazi women with the specified family history who are carriers. These probabilities need to be communicated as such, for example using phrases like 'among women with a family history comparable to yours in all important details, $\pi \times 100\%$ of women carry a deleterious allele'.

In current genetic counseling practice, a single risk estimate is often quoted to a family rather than a range of risks. Such point estimates are predicated on knowing basic parameters such as prevalences and penetrances, when there may be considerable uncertainty about them (Lange, 1986; Leal and Ott, 1995; Struewing *et al.*, 1997). This uncertainty is incorporated in the model via probabilistic sensitivity analysis, which will be discussed in Chapter 3.

1.3.7 Sequential use of Bayes' rule

One of the reasons why it is considered important to ascertain family history in genetic counseling is to assist in interpreting the results of genetic testing for individuals who elect to be tested. Genetic tests are usually accurate, but not perfect. For example, in testing for the BRCA genes there are several technologies available, most of which have a specificity near one, but can miss some deleterious mutations. Depending on the technology, sensitivity ranges from about 0.8 to about 0.98. In this context a negative result provides a Bayes factor of $(1 - \beta)/\alpha$ that ranges from $1/5$ ($\beta = 0.8$) to $1/50$ ($\beta = 0.98$). These orders of magnitude are similar to the reciprocals of those of the Bayes factors for family histories in Table 1.4, emphasizing the importance of family history information.

To fix ideas, consider a test with specificity $\alpha = 1$ and sensitivity $\beta = 0.9$. Suppose that Anne, the woman of pedigree A, elects to be tested and tests negative. Let us denote this outcome by T−, as in the liver scan example. What is the probability that Ann is a carrier? Formally, we need to compute

$$\pi(g_1 = 1 \mid x_1 > 40, x_2 = 40, \text{T}-).$$

To proceed we assume that the test outcome is conditionally independent of the family history given the genotype. This requires that the true positive rate is the same no matter what the family history is, or in other words, that the reasons why a genetic test may miss an existing deleterious mutation are unrelated to the presence of cancer in the family – an assumption which seems plausible. Then the posterior odds of being a carrier can be factored as:

$$\frac{\pi(g_1 = 1 \mid x_1 > 40, x_2 = 40, \text{T}-)}{\pi(g_1 = 0 \mid x_1 > 40, x_2 = 40, \text{T}-)}$$
$$= \frac{\pi(g_1 = 1)}{\pi(g_1 = 0)} \; \frac{p(x_1 > 40, x_2 = 40 \mid g_1 = 1)}{p(x_1 > 40, x_2 = 40 \mid g_1 = 0)} \; \frac{1 - \beta}{\alpha}$$
$$= 0.0263 \times 10.64 \times \frac{1}{10} = 0.028. \tag{1.14}$$

The first two terms in the product are the same as those of our analysis of pedigree A in Section 1.3.6. The family history of pedigree A provides evidence in favor of the mutation which is of about the same weight as the evidence against a mutation provided by a negative test.

Expression (1.14) illustrates that, formally, the product 0.0263×10.64 can be though of in two ways: as the posterior odds after we acquired the information about family history; and as the prior odds for incorporating the information about the negative test. The posterior probability $\pi(g_1 = 1 \mid x_1 > 40, x_2 = 40, \text{T}-)$ can therefore be obtained directly by (1.14) or by sequentially applying Bayes' rule, first computing the posterior probability based on family history, or $\pi(g_1 = 1 \mid x_1 > 40, x_2 = 40)$, and then using the result as the prior in expression (1.3). Expression (1.14) also illustrates that the order in which we acquire the evidence from the test and the family history is not influencing the final result, so we could reverse the order of the application of Bayes' rule.

This illustrates a general property of Bayes' rule. Suppose hypothesis H is to be revised in the light of new evidence E consisting of two empirical observations E_1 and E_2, say the family history and the genetic test of the previous example. Suppose also that the likelihood of the evidence E factors as

$$P(E_1, E_2 \mid H) = P(E_1 \mid H) P(E_2 \mid H).$$

this corresponds to the conditional independence of family history and the genetic test discussed above. In this context we can use Bayes' rule to iteratively update information by

$$P(H \mid E_1) = \frac{P(H) P(E_1 \mid H)}{P(H{=}\text{true}) P(E_1 \mid H{=}\text{true}) + P(H{=}\text{false}) P(E_1 \mid H{=}\text{false})}$$

$$P(H \mid E_1, E_2) = \frac{P(H \mid E_1) P(E_2 \mid H)}{P(H{=}\text{true} \mid E_1) P(E_2 \mid H{=}\text{true}) + P(H{=}\text{false} \mid E_1) P(E_2 \mid H{=}\text{false})}.$$

Here the output of the first application of Bayes rule is the input in the second application. This property is sometimes described by saying that 'today's posterior is tomorrow's prior'.

Conditional independence of empirical findings given the hypothesis of interest is not always satisfied in medical applications. For example, Weinstein *et al.* (1979) show, in diagnosing coronary artery disease (H), that the true positive rate from exercise electrocardiogram tests (E_2) varied across subgroups of patients classified according to type of chest pain (E_1). Subgroups at high risk because of the type of chest pain also had a higher true positive rate. Examples of the sequential use of Bayes' rule to interpret the results of several clinical findings are also discussed by Gorry and Barnett (1968) and Sox *et al.* (1988).

The discussion of pedigree A in Section 1.3.6 can also be viewed as an example of multiple findings by thinking about $x_1 > 40$ as E_1 and $x_2 = 40$ as E_2. However, these could not be considered independent conditionally on the hypothesis $g_1 = 1$ only. In fact we argued that we needed to condition on all genotypes on the pedigree before independence could be more realistically assumed. In Section 1.3.6 we also saw that conditional independence of empirical findings is not necessary for the application of Bayes' rule: it is only necessary for simple sequential updating as in (1.14).

1.4   Estimating sensitivity and specificity

In this section we reconsider the liver scan example of Section 1.2. We abandon the assumption that the sensitivity and specificity of the scan are known with certainty, and turn to the problem of treating them as unknown and estimating them based on data such as those of Table 1.1. So far we have used Bayes' rule to revise the probabilities of unknowns about patients in the light of new empirical observations, either about test results or family history. In this section we will extend this approach to unknowns that refer to populations of patients, such as α and β. The sample of Table 1.1 is to these unknowns what a family history is to the patient's genotype.

1.4.1   Population and statistical model

To keep matters simple, we will start by focusing on inference about sensitivity only. Defining the unknown in this case requires thinking about the population of all patients that are eligible for a liver scan and have liver disease, and asking what fraction of these patients would have a positive scan. This true positive rate, defined at the population level, will be our unknown, or parameter, β. The relevant sample for studying this population is the left-hand column in Table 1.1.

The population of all eligible patients is large, and it includes future patients. Therefore β is not directly observable in practice. However, a sample of patients from this population is typically very instructive about the probable values of β. The connection between the sample and the population is given by a so-called statistical model, that is, a description of the sampling mechanism and of the distribution of measurements in the population such that we can derive the probabilities of various samples conditional on a given population. In our case, conditioning on the population means simply conditioning on a hypothetical value of β, as that is the only characteristic of the population of patients with liver disease that we need to know to determine the probabilities of various samples. As in Section 1.2, we will use the term 'likelihood' to indicate probabilities of empirical evidence given unknowns of interest, a much more specific connotation than in common parlance.

The likelihood function provides the formal connection between the sample and the population unknowns. To make progress in defining it we need additional notation. We have $n = 258$ patients patient with liver disease in our sample. For generic patient i, define

$$x_i = \begin{cases} 1 \text{ if the liver scan is positive,} \\ 0 \text{ if the liver scan is negative.} \end{cases}$$

We assume that each patient is drawn at random from the population of patients with liver disease. By this we mean that the patient is selected in a way that does not bias the result of the test toward a positive or a negative scan, compared to the rest of the population. If this is the case, then

$$p(x_i = 1|\beta) = \beta,$$

that is, if the population has a true positive rate of 0.9, then the probability that a randomly drawn diseased patient will have a positive scan is also 0.9. Conversely,

$$p(x_i = 0|\beta) = 1 - \beta.$$

Using a slightly more general notation, we can express these two facts as a probability distribution of x_i, that is a list of the probabilities of all possible values of x_i. Formally:

$$p(x_i|\beta) = \beta^{x_i}(1 - \beta)^{1-x_i}, \qquad x_i = 0, 1. \tag{1.15}$$

A fundamental tool in constructing statistical models is conditional independence, encountered in Section 1.3.5. In this case we will condition on the population's true positive rate β and assume that, given β, the outcome of the test on one patient will not inform us about the probability of the outcome of the test on other patients. Mathematically, statistical independence between two variables requires that their joint probability distribution is the product of their marginals. Conditional independence is the same requirement, but applied to distributions that are conditional on some third variable or parameter, in our case β. So if we have a sample of two patients we can express conditional independence mathematically as follows:

$$p(x_1, x_2|\beta) = p(x_1|\beta)\, p(x_2|\beta).$$

More generally, if we have n patients, we can use conditional independence to build the probability distribution $p(x_1, \ldots, x_n | \beta)$ for the whole sample, starting from distributions $p(x_i | \beta)$ for the individual observations, as

$$p(x_1, \ldots, x_n | \beta) = p(x_1 | \beta) \cdots p(x_n | \beta). \tag{1.16}$$

A sample of conditionally independent random variables is often called random. It is also a special case of a so-called exchangeable sample (see Bernardo and Smith, 1994, for more details), that is, a sample such that we could lose track of the patients' order in entering the sample without losing any useful information. The distribution $p(x_1, \ldots, x_n | \beta)$ is often called the *likelihood function*, whether or not it is derived based on conditional independence. The set of assumptions used to derive the likelihood function is the statistical model.

Returning to the liver scan example, the statistical model consists of distribution (1.15) and the assumption of conditional independence. The likelihood function is

$$p(x_1, \ldots, x_n | \beta) = p(x_1 | \beta), \ldots, p(x_n | \beta) \tag{1.17}$$
$$= \beta^{x_1}(1 - \beta)^{1-x_1} \cdots \beta^{x_n}(1 - \beta)^{1-x_n} = \beta^{x_1 + \cdots + x_n}(1 - \beta)^{n - (x_1 + \cdots + x_n)}$$
$$= \beta^{231}(1 - \beta)^{27}. \tag{1.18}$$

This expression summarizes the information provided by the sample about β. It plays the same role as the probability of family history given unknown genotype in the genetic counseling example.

1.4.2   Prior and posterior distributions

Having observed the sample of Table 1.1, all possible values of β are still possible, but some will be more plausible than others. For example, values of β around 0.9 would commonly generate 231 positives out of 258, while values of β around 0.5 would only lead to 231 positives out of 258 in rare samples. Our next step is to derive $p(\beta | x_1, \ldots, x_n)$ to represent what is known about which values of β are plausible and which are not in the light of the sample. This distribution is shown, for both α and β, in Figure 1.7. The distribution of β (on the right) concentrates around the empirical proportion of positive patients in the sample, with some spread reflecting the remaining uncertainty about the population value of β.

Representing knowledge, and conversely uncertainty, about β using probabilities is intuitive and will bring a concrete contribution to the solution of complex modeling problems, as illustrated in the case studies in Part II of the book. However, to speak of a probability distribution for β we need to imagine a metapopulation, or a universe of possible populations, each with a different β. In the genetic counseling example the unknown genotype was assigned a probability distribution that represented genotype variation across

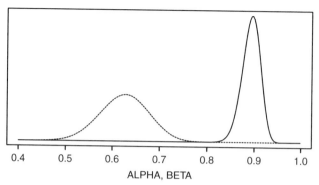

Figure 1.7   Posterior probability distributions of α (left) and β (right) based on the sample of Table 1.1, assuming a uniform prior distribution.

a concrete population of individuals. Now that we are making inferences about sensitivity (a population parameter), there is still only one population of individuals, and the probability distribution on β describes variation in the imaginary metapopulation. To many, the fact that the metapopulation is only imaginary is a sufficient reason to back away from the use of probability to represent knowledge about β. The philosophical debate about this issue is extensive and important (Barnett, 1982), but we will not discuss it in detail.

One well-known difficulty associated with the metapopulation idea is that it is not as conceptually straightforward as it was in a real population to define the prior probability, that is, the probability distribution of the possible values of β irrespective of sampling results. There are at least two approaches. The first approach is to take the prior as summarizing all scientific knowledge available about β before collecting the sample under investigation; this approach is borne out of the view that the variation in the metapopulation represents a scientist's own uncertainty about the state of the world, a view which can be traced back to the subjectivist school of probability (Ramsey, 1931; de Finetti, 1937). In medical decision making this view is especially fruitful when we need to use probability to integrate diverse and complex sources of information in modeling, or when we are interested in providing a quantitative backbone to the type of judgment that good clinicians apply when integrating facts from the medical literature. One example is the axillary lymph node application of Chapter 5.

The second approach is to take the prior as saying that the possible populations in the universe are all treated on equal footing. This approach is borne out of the view that the variation in the metapopulation represents the scientific community's standard for representing uncertainty about the state of the world, and can be traced back to seminal ideas of Jeffreys (1961), and further back to Bayes and Laplace. Priors so obtained are sometimes denoted as *reference*, or *noninformative*. In medical decision making this approach is especially fruitful when we are interested in making conclusions that depend exclusively on the experimental evidence to hand, as is often the case in regulatory settings.

The two approaches are often characterized as *subjective* versus *objective*, a characterization that overstates the differences: all statistical approaches inevitably contain important elements of subjective scientific judgment, and all typically strive to make the best use of objective empirical evidence. Because of the difficulties and potential ambiguities in defining rigorously what 'equal footing' means in most real-life situations, and because of the need of expert-based analyses to have impact and acceptance across the scientific community, most real-life applications will borrow elements from both these approaches.

Returning to the liver scan sensitivity example, we illustrate both approaches. We start with the second, and consider the first in the next subsection. In this application, we have a simple and intuitive option for treating all βs on an equal footing: β can take any value between 0 and 1, and we can assume that all these values are a priori equally likely. Although this is not the only plausible way to define 'equal footing' even in such a basic problem (Jeffreys 1961; Yang and Berger, 1997), it is simple to interpret and to justify.

An interesting justification of the use of uniform priors is provided, among others, by Edwards *et al.* (1963), who discuss the so-called *principle of stable estimation*. In their words:

> To ignore the departures from uniformity it suffices that your actual prior density change gently in the region favored by the data and not itself too strongly favor some other region. But what is meant by 'gently', by 'region favored by the data', by 'region favored by the prior distribution,' and by two distributions being approximately the same? Such questions do not have ultimate answers ...

Their paper discusses in more detail a proposal for formal conditions for proceeding as though the prior were uniform.

In the remainder of this section we will discuss in detail how to derive the posterior distribution of Figure 1.7. We will proceed under the assumption that the prior distribution is uniform, that is,

$$p(\beta) = \begin{cases} 1 \text{ if } 0 < \beta < 1, \\ 0 \text{ otherwise.} \end{cases} \tag{1.19}$$

Because we are entertaining infinitely many possible values for β, namely all values in the interval $(0, 1)$, p is no longer a list of probabilities, but a probability density function (see DeGroot, 1986, for details).

Using $p(\beta)$, $p(x_1, \ldots, x_n | \beta)$, and Bayes' rule for density functions, we can now derive $p(\beta | x_1, \ldots, x_n)$. As in the diagnostic example of Section 1.2, Bayes' rule follows directly from the definition of conditional probability:

$$p(\beta|x_1,\ldots,x_n) = \frac{p(\beta,x_1,\ldots,x_n)}{p(x_1,\ldots,x_n)} \tag{1.20}$$

$$= \frac{p(\beta)p(x_1,\ldots,x_n|\beta)}{p(x_1,\ldots,x_n)}$$

$$= \frac{p(\beta)p(x_1,\ldots,x_n|\beta)}{\int_0^1 p(\beta)p(x_1,\ldots,x_n|\beta)d\beta}.$$

The calculation in the denominator is useful for expressing the marginal distribution of the observed sample in terms of the two available distributions $p(\beta)$ and $p(x_1,\ldots,x_n|\beta)$. It corresponds to summing over all possible values of the disease status in (1.1), with the integral replacing the summation because β is continuous.

Expression (1.20) gives the probability density of the parameter β, given the data provided by the experiment or posterior density. It parallels the positive predictive value of Section 1.2 and the posterior carrier probability of Section 1.3, except for the technical difference that there are now infinitely may possible values for the unknown of interest, rather than just two. The posterior density summarizes our updated knowledge about the parameter, and can be used to make decisions, to answer scientific questions, or to make predictions about outcomes of future experiments. Inserting the results of the sample into (1.20), we have

$$p(\beta|x_1,\ldots,x_n) = 1 \cdot \frac{\beta^{231}(1-\beta)^{27}}{\int_0^1 \beta^{231}(1-\beta)^{27}d\beta}$$

$$= \frac{258!}{231!\,27!}\beta^{231}(1-\beta)^{27}. \tag{1.21}$$

To learn more about how to work out the integral in the denominator you can check your favorite book on probability (e.g. DeGroot, 1986).

Figure 1.7 illustrates the posterior probability distribution (1.21) of β. Revising the probability density function of β in the light of the sample in the left-hand column of Table 1.1 has led from a uniform (flat) prior distribution to a posterior distribution that concentrates around the empirical value of the sensitivity, with some spread reflecting the remaining uncertainty about the population value of β.

By examining the area under the posterior distribution we can address, for example, the question of the probability that the population-level sensitivity is smaller than 0.85. This is given by

$$\int_0^{0.85} \frac{258!}{231!\,27!}\beta^{231}(1-\beta)^{27}d\beta = 0.0204.$$

This calculation cannot be done in closed form, but is available in most statistical packages. When the parameter is continuous, it is common to summarize the posterior distribution with statements of the kind: β *lies between β_{low} and β_{high} with probability 0.95, or 0.99.* This has the advantage of conveying information about our best guess of the unknown as well as the accuracy of that guess. Moreover, this is done in a way which is easily understood by nonexperts. One way of generating an interval that contains β with probability .95 is by determining β_{low} and β_{high} so that

$$P(\beta < \beta_{\text{low}}) = \frac{0.5}{2} \quad \text{and} \quad P(\beta > \beta_{\text{high}}) = 1 - \frac{0.5}{2}.$$

This procedure leads to the so-called equal-tailed posterior probability intervals. Discussion of various alternative approaches for summarizing a distribution using an interval is given in Gelman *et al.* (1995).

Figure 1.7 also illustrates the posterior probability distribution (1.21) of α. Using the right-hand column of Table 1.1, and the same line of argument used to derive the posterior distribution on β, we obtain that

$$p(\alpha|x_1, \ldots, x_n) = \frac{86!}{54!\, 32!} \alpha^{54}(1 - \alpha)^{32}, \tag{1.22}$$

the density function that is depicted on the left in Figure 1.7. Again, the distribution concentrates around the empirical estimate of specificity. Compared to the distribution of β, it is more dispersed, reflecting the smaller sample size.

The argument used to derive expressions (1.21) and (1.22) goes back as far as Bayes himself, who was concerned with this very problem in his famous essay (Bayes, 1763). Stigler (1986) gives a careful account of the early developments of the idea of 'inverse probability', that is, the problem of converting probabilities of samples given hypotheses into probabilities of hypotheses given samples.

1.4.3 Distributions of positive and negative predictive values

In Section 1.2 we considered the question of how the result of a liver scan should assist in the clinical management of a patient, and addressed it by considering the probability π^+ that the patient testing positive is actually diseased, or the positive predictive value. We derived an expression for π^+ as a function of α and β,

$$\pi^+ = \frac{\pi\beta}{\pi\beta + (1 - \pi)(1 - \alpha)},$$

and evaluated it for specific choices of α and β. Now we are no longer assuming we know α and β with certainty, and we represent our uncertainty via a probability distribution. These distributions imply a distribution on π^+ as well. One way to think about this distribution is as follows. Imagine drawing

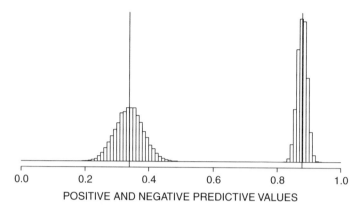

0.0   0.2   0.4   0.6   0.8   1.0
POSITIVE AND NEGATIVE PREDICTIVE VALUES

Figure 1.8 Histograms of 1000 sampled values of the positive predictive value π^+ (right) and the probability π^- of disease given a negative test (left) based on samples from the posterior distribution of α and β.

a sample of values of α and β from their respective metapopulation, assuming that α and β are distributed in the metapopulation according to the posterior distribution of Figure 1.7. Each pair of α and β determines a π^+, so that the sample of αs and βs determines a sample of π^+s. Figure 1.8 depicts one such sample, obtained using a computer simulation, for both π^+ and π^-. In Chapter 3 we will discuss in more detail how to obtain samples of simulated values from distributions like those of Figure 1.8. In this example the simulation is a shortcut for performing the mathematical derivation necessary for working out the distributions of π^+ and π^-.

Both the distributions of Figure 1.7 and those of Figure 1.8 quantify the limitations of our knowledge about the performance of hepatic scintigraphy. But because π^+ and π^- are the quantities that are most directly relevant clinically, Figure 1.8 represents these limitations in a way that is more directly useful. For example, it says that among patients who tested negative for the liver scan, the fraction of patients that are diseased could be anywhere between about 0.2 and about 0.45. It also emphasizes that the limitations of our understanding of the test are more important for patients testing negative than they are for those testing positive, whose predictive value can be estimated with greater accuracy. This simple simulation exemplifies one of the strengths of probabilistic inference: the ease of translating uncertainty about population parameters into uncertainty about prediction of patient outcomes.

1.4.4 Beta prior distributions

As an alternative to specifying priors that treat all possible values of the parameter on an equal footing, one can use the prior distribution to incorporate knowledge derived from sources other than the sample. Often this is done via parametric families that contain a wide variety of shapes, but also lend themselves to simple mathematical treatment. In the case of binary data, the most popular choice of prior is the beta distribution (DeGroot, 1986; Berry, 1996) – a name which is unrelated to our investigating a parameter called β. For positive numbers a and b the beta distribution with parameters a and b, denoted by Beta(a, b), has density function

$$p(\beta) = \frac{1}{B(a,b)} \beta^{a-1}(1-\beta)^{b-1} \tag{1.23}$$

where $B(a, b) = \int_0^1 \beta^{a-1}(1-\beta)^{b-1}d\beta$. If a and b are integers, then

$$B(a, b) = \frac{(a-1)!(b-1)!}{(a+b-1)!}.$$

By choosing a and b we can represent many different situations. For example, if $a = b = 1$, $p(\beta)$ is the uniform distribution of the previous section. The distributions of Figure 1.7 is also a special case. As can be seen by checking equations (1.21) and (1.22), the distribution of β is Beta$(232, 28)$, while the distribution of α is Beta$(55, 33)$.

There are several ways of interpreting and choosing the values of a and b in a beta distribution. One is by understanding how to control the location and the shape of the distribution. The mean of the beta distribution (1.23) is

$$E(\beta) = \frac{a}{a+b},$$

so that the relative magnitudes of a and b control the location of the distribution: when a is large compared to b the mean will be close to 1. The standard deviation is

$$\sqrt{\frac{ab}{(a+b+1)(a+b)^2}} = \sqrt{\frac{E(\beta)[1 - E(\beta)]}{a+b+1}},$$

so that for fixed mean the sum of a and b determines the spread of the distribution.

One reason why the beta distribution is popular is that it leads to a very simple rule for updating the distribution of β after a sample of conditionally independent individuals with outcomes $x_1, \ldots, x_n$ is observed. In particular, if we define $y = x_1 + \cdots + x_n$, the posterior density function is given by:

$$
\begin{aligned}
p(\beta|x_1, \ldots, x_n) &= \frac{p(\beta)p(x_1, \ldots, x_n|\beta)}{p(x_1, \ldots, x_n)} \\
&= \frac{\beta^{a-1}(1-\beta)^{b-1}\beta^y(1-\beta)^{n-y}}{B(a,b)\int_0^1 \frac{1}{B(a,b)}\beta^{a-1}(1-\beta)^{b-1}\beta^y(1-\beta)^{n-y}d\beta} \\
&= \frac{1}{B(a+y, b+n-y)}\beta^{a+y-1}(1-\beta)^{b+n-y-1}.
\end{aligned}
$$

We have just proved that if the prior distribution is Beta(a, b) then the posterior distribution is again in the beta family, and specifically is Beta$(a + y, b + n - y)$. Additional material on this family and its role in Bayesian inference for binary data can be found in Berry (1996) and Gelman *et al.* (1995).

Updating leads to a new distribution but not to a new type of distribution, a fact which makes the beta a *conjugate* distribution to the likelihood function for binary observations. A comprehensive study of conjugate prior distributions is given by Raiffa and Schleifer (1961). General theoretical considerations, interpretation, and characterization of situations in which conjugate priors are available are discussed by Diaconis and Ylvisaker (1979).

A Beta$(a + y, b + n - y)$ posterior can occur in many ways: one is by specifying a Beta(a, b) and observing y positive out of n patients. Another is by having a uniform prior and observing $a + y$ positive out of $a + b + n$ patients. This suggests another way of specifying the parameters a and b when choosing a beta prior distribution. Specifically, we can think of the Beta(a, b) distribution as incorporating the evidence from a sample of $a + b$ conditionally independent patients from the same population from which the actual sample is drawn, when a of those patients are positive. This has two implications. First, the prior can be used to incorporate evidence from existing relevant experiments; second, even when such prior evidence is not available, the prior can be specified and interpreted using a 'prior sample' interpretation to facilitate the specification of a and b.

Figure 1.9 shows five beta distributions with various choices of a and b. These can be interpreted as possible priors or as the posterior distributions that correspond to starting with a uniform prior and observing the following samples:

Posterior density	Sample
0	
1	$x_1 = 1$
2	$x_1 = 1, x_2 = 1$
3	$x_1 = 1, x_2 = 1, x_3 = 0$
4	$x_1 = 1, x_2 = 1, x_3 = 0, x_4 = 1$

As the sample gets larger the posterior distribution gets more concentrated around values close to the proportion of positive results in the sample.

1.4.5 Prior-data conflict

The 'prior sample' interpretation of the beta prior is appealing and can substantially simplify both prior specification and inference. However, it assumes that the patients in the 'prior sample' are drawn from a population

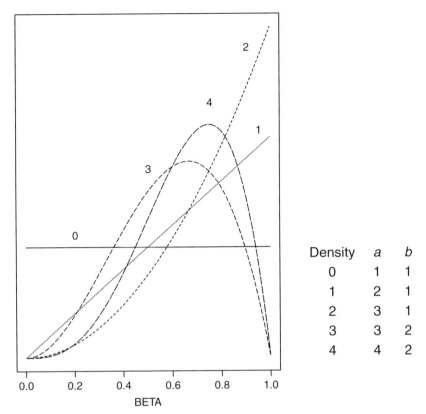

Density	a	b
0	1	1
1	2	1
2	3	1
3	3	2
4	4	2

Figure 1.9   A sequence of beta distributions with increasing values of $a + b$.

that is identical to that of the patients in the actual sample. This feature may not always reflect practical situations.

To illustrate this point we revisit the inference on α described in Figure 1.7 under a different prior. We will imagine that the prior is elicited based on an expert's previous experience with liver scans. The expert considers his/her experience equivalent to having observed about 100 healthy patients, about 90% of whom have negative liver scans. The resulting posterior distribution is shown in the top left-hand panel of Figure 1.10: its location is intermediate between the prior and the normalized likelihood – which is also the posterior under the uniform prior of the previous section. There is little overlap with either of the 95% probability regions. The spread of the posterior is smaller than both the prior's and the posterior's.

This is sensible if one can confidently consider the expert's experience and the experiment to be drawing from identical populations. But this may not always be the case, either because of peculiarities of the study design, or because of biases in the expert's experience. More generally, in science, while

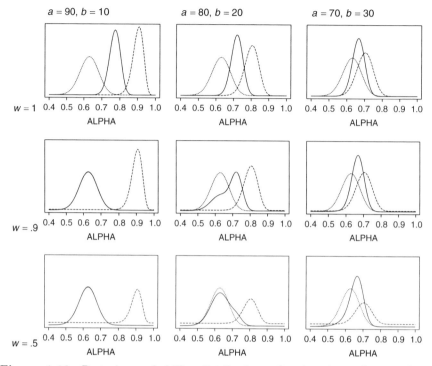

Figure 1.10 Posterior probability distributions of α based on the sample of Table 1.1 under nine different choices of the prior distribution (1.24). Continuous lines are posterior distributions, dashed lines are priors, and dotted lines are the normalized likelihood (also corresponding to the posterior distribution with a uniform prior.)

consistent findings tend to reinforce each other, discordant ones seldom lend support to compromise positions that are inconsistent with both of the initial findings.

A more general approach to prior elicitation is based on the idea of introducing additional uncertainty about whether or not the two pools of patients are indeed identical. This generates distributions that combine prior and sample information when the two do not appear to be in conflict, but discounts the prior information when there is apparent conflict between prior and data. Technically one such prior can be specified by assigning a weight w to the two populations being identical and a weight $1 - w$ to the two populations being so different that inference should be based solely on the sample. Mathematically

$$p(\alpha) = \left[(1 - w) + wB(a, b)\alpha^{a-1}(1 - \alpha)^{b-1} \right], \qquad (1.24)$$

with $0 \le w \le 1$. We will call this a *mixture* prior, because it is a composite of a Beta (a, b) and a uniform. The conjugate prior is a special case occurring when either $w = 1$ or $a = b = 1$.

Priors of this type are depicted in Figure 1.10. When $w < 1$ (i.e. in the six panels at the bottom), priors are characterized by the combination of a hump corresponding to the expert's experience and a uniform layer spread over the region that the expert did not consider a priori plausible. This layer guarantees a healthy measure of open-mindedness in the learning process.

Figure 1.10 contrasts the effects of the conjugate prior and the mixture prior. We have $a + b = 100$ throughout. The left-hand column considers the expert's experience to be such that about 90% of healthy patients would have negative liver scans. This is in conflict with the results of the study, in which less than 65% of healthy patients had negative liver scans. We have noted how the conjugate prior leads to the conclusions that the likely αs for the population are between those of the expert and the data; and b) the accuracy of the conclusions is higher than it would be if based on just the prior or just the data. (The latter is taken to mean a uniform prior.) By contrast, the mixture priors of the bottom two rows, based on $w = 0.9$ and $w = 0.5$, both lead to inferences that coincide with those based on uniform priors. The conclusions are that the likely αs for the population are those supported by the data; and the accuracy of the conclusions is the same as it would have been if we had had a uniform prior. The expert's experience is effectively ignored because it is in too great a conflict with the data.

When is conflict 'too great'? In our simple situation, we can look at other combinations of w, a and b to get a sense for this. In the middle and right-hand columns of Figure 1.10, the proportion of healthy patients who would have negative liver scans in the expert's experience is decreased to 80% and to 70%, respectively. The effect of this depends on the value of w. In the bottom row, where $w = 0.5$, the 80% prior experience is still largely discounted as too discordant. There is only the shadow of a doubt being raised, as reflected by the right-hand tail of the posterior and the slight increase in the spread. On the other hand, the 70% experience is not discounted and the posterior cumulates the accuracy of the two samples, with a spread that is smaller that both the prior and the posterior.

Discounting of discordant priors is less pronounced in the middle row, corresponding to $w = 0.9$. Comparison of the middle column in the two bottom rows illustrates that the mode of the posterior distribution can be sensitive to the choice of w. However, the 95% credible interval is much less sensitive, and in both cases regions with high likelihood are included in the credible region while regions with high prior might be excluded.

1.4.6 Future patients

One of the most far-reaching practical implications of using probability to quantify uncertainty about parameters such as α and β is that it becomes straightforward to make predictions about future samples of patients in a way that recognizes the parameter uncertainty. One example where this is important is in making predictions about the outcomes of future samples of patients. This is useful in planning additional experiments or in deciding whether to stop or continue an ongoing study.

To illustrate this mechanism, let us begin with the case in which we have not observed anything and we are interested in the outcome of the liver scan in the first patient with liver disease. Also, let us assume we have a uniform prior on β. Predictions about x_1 are based on the (marginal) distribution of x_1, which is found from the relationship

$$p(x_1) = \int_0^1 p(x_1|\beta)\, p(\beta)\, d\beta.$$

This can be thought of as follows: we first draw a population (i.e. a value of β) from the imaginary metapopulation, using the prior distribution. Then we focus on that population and draw a patient. The probability of ending up with a particular sequence of β and x_1 is $p(x_1|\beta)p(\beta)$. If we are interested in the probability of x_1 irrespective of the metapopulation that led to it, than we must recognize that the same value of x_1 could have been obtained under different metapopulations, and sum (or integrate) over all possible paths that lead to that x_1. Because we are using a uniform prior,

$$p(x_1|\beta)\, p(\beta) = \beta^{x_1}(1-\beta)^{1-x_1}$$

and

$$p(x_1 = 1) = \int_0^1 \beta\, d\beta = \frac{1}{2},$$

$$p(x_1 = 0) = \int_0^1 (1-\beta)\, d\beta = \frac{1}{2}.$$

Now suppose that we have observed a positive liver scan result in the first patient, $x_1 = 1$, and want to predict the outcome x_2 of the liver scan in patient 2. We need $p(x_2|x_1)$. The general formula is:

$$
\begin{aligned}
p(x_2|x_1) &= \int_0^1 p(x_2, \beta|x_1)\, d\beta \\
&= \int_0^1 p(x_2|\beta, x_1)p(\beta|x_1)\, d\beta \qquad (1.25) \\
&= \int_0^1 p(x_2|\beta)p(\beta|x_1)\, d\beta.
\end{aligned}
$$

The first equality expresses the marginal density of x_2 by integrating the joint density of x_2 and β; this time all the probability density functions are conditional on x_1. The second equality comes from writing the joint density of β and x_2 as the product of a density for β and a conditional density for x_2 given β as we did in the prediction for patient 1; again, all the densities are conditional on x_1. The third equality comes from the conditional independence of x_1 and x_2 given β via the intermediate steps $p(x_2|\beta, x_1) = p(x_1, x_2|\beta)/p(x_1|\beta) = p(x_1|\beta)p(x_2|\beta)/p(x_1|\beta)$. We could have arrived at the middle line of (1.25) directly by noticing that it is just the same as the formula for $p(x_1)$ but with x_2 replacing x_1 and the current density for β, $p(\beta|x_1)$, replacing the previous density for β, $p(\beta)$.

Let us now go back to our discussion and turn to the predictive probability of a positive liver test in patient 2. In our example, we have

$$p(\beta|x_1 = 1) = 2\beta,$$

as you can check by using an argument like the one that led to (1.17). Observing that $x_1 = 1$ has shifted the mass of the distribution toward higher values of β, as one would expect. Now in making a prediction for the next observation we incorporate the new information that higher values of β are more likely than lower values and, via (1.25), obtain that

$$p(x_2 = 0|x_1 = 1) = \int_0^1 (1 - \beta)\, 2\beta \, d\beta = \frac{1}{3},$$

$$p(x_2 = 1|x_1 = 1) = \int_0^1 \beta \, 2\beta \, d\beta = \frac{2}{3}.$$

Similar calculations would lead to $p(x_2 = 0|x_1 = 0) = 2/3$ and $p(x_2 = 1|x_1 = 0) = 1/3$.

These simple calculations illustrate how past observations can be incorporated in making predictions about future observations, and how they affect the distribution of the future observations. Because patient 1 teaches us about patient 2, the two outcomes are not independent, as illustrated by the fact that the distribution of x_2 depends on the outcome of x_1. This is not in conflict with the fact that they are conditionally independent given β. Conditional independence only implies that all that we learn from patient 1 about patient 2 goes through β. These equations emphasize that to compute the predictive distribution we revise knowledge about β using the observed data and Bayes' rule and use the updated distribution to make predictions about future data. The distribution on the parameter acts as an intermediary between past and future data, condensing all that is relevant about the past in making predictions about the future.

More generally, if we have observed a random sample of n observations and we are interested in predicting the next, we use:

$$p(x_{n+1} \mid x_1, \ldots, x_n) = \int_0^1 p(x_{n+1}|\beta)p(\beta|x_1, \ldots, x_n)d\beta. \qquad (1.26)$$

leading to

$$P(x_{n+1} = 1|x_1, \ldots, x_n) = \frac{y+1}{n+2},$$

where $y = \sum_{i=1}^n x_i$. For example, the probability that an additional patient drawn from the diseased population from which the left-hand column of Table 1.1 was drawn is tested positive is $(231 + 1)/(258 + 2) = 0.892$.

To derive this statement, observe that under a uniform prior, the right-hand side of (1.26) is the mean of a Beta $(y + 1, n - y + 1)$ distribution. If instead of a uniform prior we were using a general Beta(a, b), the probability of a positive liver scan would be

$$P(x_{n+1} = 1|x_1, \ldots, x_n) = \frac{y+a}{n+a+b}.$$

1.5   Chronic disease modeling

1.5.1   A simple Markov process for strokes – the μSPPM

We now turn to an example concerning estimation of the parameters of a model that describes the history of patients at high risk of stroke. We introduce a simple model which we term the micro Stroke Prevention Policy Model, or μSPPM, as it is modeled after the Duke PORT's Stroke Prevention Policy Model (Matchar et al., 1993, 1997), a simulation model whose goal is to predict the long-term health outcomes and costs of patients at risk of stroke, under alternative prevention and/or treatment strategies. We will return to the μSPPM on several occasions in Chapters 2 and 3.

The core of the μSPPM is a description of the natural histories of stroke-related events in a cohort of individuals. Figure 1.11 (Parmigiani et al., 1997) illustrates the possible histories. Health states, represented by the ovals in the figure, are HEALTH (abbreviated H), TIA (transient ischemic attack, abbreviated T), STROKE (S), and DEATH (D). Being in the S state means having suffered a stroke at some point in the past. The actual event of a stroke is the transition from H or T to S. Transitions between states are represented by the arrows. Because of the way states are defined, transitions are only possible in one direction. The resulting model is called a stochastic compartment model (Manton and Stallard, 1988).

Next to each arrow in Figure 1.11 is the corresponding transition rate, that is, the average number of transitions per unit of time. These transition rates

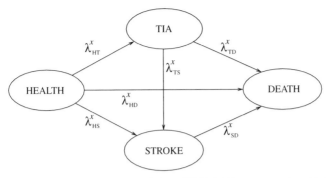

Figure 1.11   States and transitions of the μSPPM. The process is a highly stylized representation of stroke histories.

are unknown and are the focus of our inference. We will allow the transition rates to depend on whether the patient has stenosis of the arteries, defined by a binary covariate x, with $x = 1$ corresponding to the presence of stenosis. To simplify, we assume that stenosis status does not change over time. We indicate by λ^x_{HT} the transition rate from H to T, conditional on covariate x, and use a similar notation for the other transition rates. Also, we will use λ for the vector of all the λs and

$$\lambda^x_{\mathrm{H}} = \lambda^x_{\mathrm{HT}} + \lambda^x_{\mathrm{HS}} + \lambda^x_{\mathrm{HD}},$$
$$\lambda^x_{\mathrm{T}} = \lambda^x_{\mathrm{TS}} + \lambda^x_{\mathrm{TD}}$$

for the transition rates out of H and out of T, respectively. To begin with a simple example, we assume that transition rates, while dependent on the to and from states, are independent of age and of the length of time already spent in the current state. This results in a Markov process. Consider state T. Each individual that reaches T will spend in it a random amount of time. The distribution of this time is exponential with mean $1/\lambda^x_{\mathrm{T}}$. Eventually, he/she will move out, to either S or D. The move will be to S with probability $\lambda^x_{\mathrm{TS}}/\lambda^x_{\mathrm{T}}$ independent of the time spent in T. Similar considerations apply to the other states. If you are not familiar with these concepts, you can consult a chapter on finite-state, continuous-time Markov processes in a book on stochastic processes such as Bhat (1984). This modeling approach is clearly too restrictive in reality. In Section 1.5.2 we consider a more general class of models.

Here we focus on inferences on the transition rates. The most appropriate type of data for estimating these parameters in a model like the μSPPM is a cohort study of healthy individuals followed over time. A patient history consists of the states visited by that patient and the corresponding times spent in the states (or sojourn times). Table 1.5 demonstrates the data format in a hypothetical example. Each row represents a transition, so there can be multiple rows for each patient. Under our simple assumptions, the

Table 1.5 Format of the hypothetical cohort study for estimating the parameters of the μSPPM. Each row corresponds to a transition.

Patient	Stenosis status	Original state	Years in state	Resulting state
1	0	HEALTH	0.83	TIA
1	0	TIA	1.46	STROKE
1	0	STROKE	0.31	DEATH
2	0	HEALTH	9.75	DEATH
3	1	HEALTH	6.99	TIA
3	1	TIA	2.42	STROKE
3	1	STROKE	0.54	DEATH
4	0	HEALTH	0.01	STROKE
4	0	STROKE	0.45	DEATH
...				
...				
998	0	HEALTH	15.41	TIA
998	0	TIA	0.37	DEATH
999	1	HEALTH	0.43	TIA
999	1	TIA	0.20	STROKE
999	1	STROKE	0.10	DEATH
1000	1	HEALTH	4.68	TIA
1000	1	TIA	0.30	STROKE
1000	1	STROKE	0.07	DEATH

transition rate between any two states can be estimated by the total number of transitions observed in the sample, divided by the time at risk in each state. As we saw, it is more interesting to provide complete probability distributions for the λs, reflecting what can be learned from the analysis of the cohort.

In the Markov model used in this section, it is simple to derive the likelihood function and the a posteriori probability density function of the transition rates. To avoid complicated notation, consider the probability of observing the path H $\to$ T $\to$ D, with sojourn times y_{H} and y_{T}. This is

$$p(y_{\mathrm{H}}|\lambda_{\mathrm{H}}^1)p(\mathrm{H} \to \mathrm{T}|\lambda_{\mathrm{H}}^1, \lambda_{\mathrm{HT}}^1)p(y_{\mathrm{T}}|\lambda_{\mathrm{T}}^1)p(\mathrm{T} \to \mathrm{D}|\lambda_{\mathrm{T}}^1, \lambda_{\mathrm{TD}}^1)$$
$$= \lambda_{\mathrm{HT}}^x e^{-\lambda_{\mathrm{HT}}^x y_{\mathrm{H}}} \lambda_{\mathrm{TD}}^x e^{-\lambda_{\mathrm{TD}}^x y_{\mathrm{T}}}. \tag{1.27}$$

The likelihood function is a product of patient-specific contributions like the one above. The components of the likelihood that refer to the individual parameters can also be factored in a product, and all have the same structure. For example, consider the H to T transition for patients with stenosis status x. Call N_{HT}^x the number of transitions from HEALTH to TIA in the subset of patients with stenosis x, and t_H^x the total time at risk in the HEALTH state in the same patients. The component of the likelihood function that includes the parameter λ_{HT}^x is

$$(\lambda_{HT}^x)^{N_{HT}^x} e^{-\lambda_{HT}^x t_H^x}.$$

If the elements of λ are a priori independent, then the posterior distribution is also a product of terms that refer to the individual parameters. To derive $p(\lambda_{HT}^x | \text{Data})$ we can, for example, assume vague initial knowledge about the unknown transition rate, modeled via a so-called noninformative prior of the form $\pi(\lambda_{HT}^x) \propto 1/\lambda_{HT}^x$, and use Bayes' rule. This leads to the following posterior probability density function of the rate λ_{HT}^x:

$$p(\lambda_{HT}^x | \text{Data}) = \frac{t_H^x}{(N_{HT}^x - 1)!} (t_H^x \lambda_{HT}^x)^{-1 + N_{HT}^x} e^{-\lambda_{HT}^x t_H^x}, \qquad (1.28)$$

a gamma distribution. Similar expressions hold for all the other rates. Alternative a priori assumptions can also be used to incorporate expert opinion or information from published studies investigating the same transition rate. A popular choice is to use a gamma prior distribution, which is the conjugate prior (Bernardo and Smith, 1994).

Using these posterior distributions and the law of total probability we can, as in Section 1.4.6, determine the predictive distribution for outcomes of future patients. For example, the probability of observing the path H $\rightarrow$ T $\rightarrow$ D, with sojourn times y_H and y_T, in a patient with stenosis is

$$\int p(y_H | \lambda_H^1) p(\text{H} \rightarrow \text{T} | \lambda_H^1, \lambda_{HT}^1) p(y_T | \lambda_T^1) p(\text{T} \rightarrow \text{D} | \lambda_T^1, \lambda_{TD}^1) p(\lambda | \text{Data}) d\lambda$$

$$= \int \lambda_{HT}^1 e^{-(\lambda_{HT}^1 + \lambda_{HT}^1 + \lambda_{HD}^1) y_H} \lambda_{TD}^1 e^{-(\lambda_{TS}^1 + \lambda_{TD}^1) y_T} p(\lambda | \text{Data}) d\lambda.$$

We obtained this using expressions (1.27) and (1.28). The integrals are six-dimensional and range over all possible values of the vector λ. It is possible to solve this integral analytically and obtain a closed-form expression for this probability. In practice, it is more convenient to analyze this distribution and its implications using simulation, as will be described in Section 3.3.2.

1.5.2   Estimating transition models from cohort data

The example of Section 1.5.1 considers a Markov process with time-independent transition rates. This model is simple to interpret and fit, but is highly unrealistic in the stroke application. For example, stroke and TIA risks in asymptomatic patients increase with age. Also TIA patients have a high risk of stroke in the months immediately following a TIA, but then their risk can revert to the levels of a high-risk individual who did not have a TIA.

This calls for time-dependent transition rates. In addition, the probability of making each of the transitions considered by the μSPPM would depend on prognostic factors.

In general, we face the problem of estimating time-dependent transition probabilities as well as selecting which covariates affect the various transition probabilities. One additional complication is that the individuals who are at high risk of TIA are also at high risk of stroke, even within subgroups with homogeneous risk factors. An outline of a methodology for this type of analysis follows. We use the notation s for a state, y for time in the current state, and x for the patient covariates. We discuss a simple approach that offers a direct way to produce transition probabilities. It is based on combining, for each state, two elements: a proportional hazard regression for the time to the next event y depending on covariates and on whether individuals have previously visited certain states; and a polytomous regression for the type of event depending on covariates, on previously visited states and on sojourn time t.

Transition probabilities can be obtained from the resulting estimates as follows. Consider an individual in state s. Let $p_s(y|x)$ be the estimated probability of a patient with covariates x remaining in state s for y units of time, and let $Q_s(t|x)$ be the state-specific survivor function for state s, or, in other words, the proportion of individuals who are still in state s after y units of time. Both of these can be derived from the proportional hazards model for the time to next event. Also, from the polytomous regression for the type of event, we can obtain $p_s(s'|y, x)$ – the probability that a transition from s occurring after a sojourn of y units will be to state s'. We have dropped the dependence on previously visited states for notational simplicity. For an individual in state s at time 0, the probability of making a transition from s to s' at time y is then $Q_s(y|x)p_s(s'|y, x)$.

An important component of this modeling strategy concerns the effect of covariates on transition probabilities. In the discussion above, covariates are generically indicated by a vector x. However, not all the covariates will generally affect all transitions. Also, when there is a large number of transitions to be estimated, it will not be possible to reliably estimate the effects of all covariates on all transitions due to sample size limitations. So some of the elements of x may be multiplied by coefficients that are zero. Choosing which coefficients should be set to zero is difficult. Standard variable selection approaches consider one relationship at the time. Here the various proportional hazards and polytomous regression are likely to be strongly interrelated, which calls for an assessment of variable selection as a whole, by comparing the resulting overall model prediction with the actual patient histories.

2

Decision making

2.1 Summary

In this chapter we present the fundamental ideas of Bayesian decision theory. In Chapter 1 we saw how uncertainty and variation can be quantified using probability. In this chapter we will discuss quantifying the value of health outcomes to individuals, using a numerical measure called utility. There are two reasons why both these quantifications are critical in medical problems. First, the consequences of alternative actions cannot generally be completely anticipated, so that a measure of the plausibility of different health outcomes is necessary in order to reach a good decision. Second, medical decisions often involve trade-offs between length and quality of life, or affect morbidity levels without ultimately affecting mortality. Health outcomes have multiple dimensions and it is rarely the case that any one of these dimensions can fully capture the objectives of medical care. Also, individuals value the same health outcomes to highly varying degrees.

The extent to which general quantitative notions of rationality, such as expected utility theory, can be brought to bear in the development of clinical decisions and policies varies widely with the specific medical problem that is being considered, the decision maker(s), the degree of conflict among alternative goals, the degree to which uncertainties and values can be reliably quantified, and so on. The strength of decision-theoretic ideas in medicine is that of providing a structure and an end to the process of gathering, organizing, and integrating the quantitative information that is relevant to a decision. In this sense, despite the limitations of its axiomatic foundations, and the cognitive difficulties with representing and communicating values and utilities to the general public, decision-oriented quantification is an almost indispensable component of good medical decision making.

Two topics are discussed in the abstract in this chapter: maximization of expected utility as a principle of rationality, and maximization of expected utility as a principle for learning from evidence. In addition, two small case studies are used to put ideas into context.

1. *Measuring the value of the outcome of a major stroke.* Health outcomes that affect quality of life over an extended period of time play a critical role in decision making for chronic diseases. Quantifying the value of avoiding such outcomes provides the basis for measuring the importance of preventive interventions. A stroke is a case in point. We use this example to illustrate three widely used methodologies for measuring the value of health outcomes: the standard gamble approach, the time trade-off approach, and a combination of the two that incorporates differential preferences for outcomes depending on how far in the future they may be. The time trade-off analysis is based on the Duke PORT study (Samsa *et al.*, 1999).

2. *A miniature stroke policy model.* In Chapter 1 we presented the μSPPM, a miniature model for prediction of the natural course of stroke-related events in the absence of intervention (Parmigiani *et al.*, 1997). Here we will expand the μSPPM so that it can consider the effects of health interventions, and use it as a tutorial example to discuss the implications of the expected utility principle in medical decision making, both from an individual and a societal perspective.

One of the earliest systematic treatments of quantitative approaches to medical decision making is Lusted (1968). Comprehensive texts include the influential work of Weinstein *et al.* (1980), the accessible introductory work of Sox *et al.* (1988), and the volume by Chapman and Sonnenberg (2000). An excellent introduction to decision analysis techniques, written primarily with business applications in mind, but relevant to health care decisions as well, is Clemen and Reilly (2001). More statistically oriented treatments are provided by Lindley (1985) and French (1986). Specific discussion of the role of utility theory can be found in Pliskin *et al.* (1980) and Torrance (1986). A closely related field is that of cost-effectiveness analysis, heralded by Weinstein and Stason (1977) and discussed in detail by Gold *et al.* (1996) and Sloan (1995).

2.2   Foundations of expected utility theory

Normative decision theory investigates decision making under uncertainty, its logic, its mathematical structure, and its implementation. In medicine, it can be used to develop general approaches to determining optimal medical choices given a certain body of evidence and values. This section presents in the abstract some of the key ideas of decision theory and its axiomatic foundations. The perspective will be that of an individual decision maker (DM), who is either a patient or acts on behalf of a patient, and is interested in determining an optimal course of action using formal modeling. Both the values and beliefs that are being modeled are the DM's own. It is critical, from the point of view of the foundations, that both the values and the beliefs are those of the same individuals. In practice, this does not require every DM to develop a comprehensive understanding of a medical field for this theory to apply. DMs can embrace their own physicians' views, or adopt the views of published results derived from comprehensive analyses.

In many applications, it is neither desirable nor practical to take the expected utility view to the letter. Sometimes, health-related decisions involve personal aspects that are difficult to quantify reliably along one-dimensional utilities. At other times, the level of detailed quantification required by decision-theoretic approaches is not achievable, either because of cognitive or psychological difficulties in relating a DM's values to quantitative utilities, or because of gaps in medical knowledge. Yet the decision approaches that emerge from the formal exercise of determining the optimal course of action, while not a replacement for human judgment, can provide valuable assistance and insight in a broad range of individual and policy decision situations.

2.2.1 Formalizing health interventions and their consequences

Health care decisions require the consideration of alternative *actions*, whose consequences, or *outcomes*, depend on circumstances that are unknown, either because they are not directly observable or because they will take place in the future. Decision theory provides a formal model for this choice based on both the desirability and likelihood of the actions' consequences.

Three critical notions in developing a formal theory are action, outcome, and parameter. Outcomes, denoted by z, are future health histories of individuals or groups, described in all aspects that are relevant enough to the decision at hand. Actions, denoted by a, are the alternatives available to the DM. An action can be a treatment, a community intervention, a prevention strategy that evolves with patients' risk factors, and so on. Actions have an effect on outcomes, but generally do not determine outcomes with certainty. If the same medical treatment is applied to a homogeneous group of patients, these will still experience different future health histories. We can describe an action by the collection of health outcomes that result from it over a cohort of patients that are similar to the DM in all relevant aspects. If we knew this cohort in full detail, then the DM's future could be described as a randomly selected outcome from the cohort. Generally it will not be possible to provide a perfect description of this cohort, but it may be possible to provide one that is based on a sample of individuals and additional expert opinion on, for example, who is similar enough to the DM for the purposes of the problem at hand. The parameters θ represent a characterization of the true, but unobserved, 'complete cohort'.

Formally, an action can be described by the set of probability distributions over outcomes that it gives under each possible parameter value. We will use the notation

$$p_a(z|\theta) \tag{2.1}$$

to indicate this. When a decision problem explicitly considers patient-specific risk factors or covariates x, an action will be defined by the probability distributions

$$p_a(z|x, \theta). \tag{2.2}$$

Throughout our discussion of foundations, we will consider (2.1) only. The sets $\mathcal{A}$, $\mathcal{Z}$, and Θ, include, respectively, all available actions, all possible consequences and all parameter values.

For example, in the μSPPM model of Section 1.5.1 there is a single action, standard care (additional alternatives will be considered later in this chapter). The outcome z is the patient's trajectory through the model of Figure 1.11, inclusive of all transitions and transition times. Each z brings with it a history of health status in terms of symptoms, functionality, ability to perform daily activities, and so on, associated with the states visited. The parameters are the true unknown transition rates from state to state. These describe properties of populations of patients and are known only up to estimation error.

Another simple example is illustrated in Table 2.1. The DM must choose between two mutually exclusive treatments, A and B. The trade-off arises because of a genetic modifier of the treatments' effects. Specifically, treatment A is effective in patients with a responsive genotype, but not effective in others, while treatment B is moderately effective in all patients. The relevant parameter θ is in this case the genotype, assumed unknown. If we knew θ we could identify the appropriate reference cohort for the patient. For each combination of treatment and genotype, we have a potentially different probability distribution of recovery times, succinctly described by adjectives in the table.

2.2.2   The expected utility principle

In the treatment choice example of Table 2.1, there is a trade-off between a normal recovery and an immediate, but uncertain recovery. Addressing this trade-off requires balancing the severity of symptoms against the probability of a responsive genotype. For example, appropriate actions may differ

Table 2.1   Tabular representation of the choice between two treatments, A and B, in the presence of a genetic modifier of the treatment effect. Columns represent unknown parameter values; entries in the table represent health outcomes. Exact recovery times are unknown. An action is characterized by a list of probability distributions over outcomes, corresponding to a different parameter values.

	Responsive genotype	*Unresponsive genotype*
Treatment A	Immediate recovery	Slow recovery
Treatment B	Normal recovery	Normal recovery

between patients who are known to carry a genetic marker linked with the responsive genotype, and others. Likewise, actions may differ for patients for whom a slow recovery could cause serious complications or psychological sequelae.

Decision-theoretic approaches quantify uncertainty over genotype using probability, and capture the severity of symptoms by means of a numerical measure called utility. For each action, the weighted average of the utility with respect to the probability is used as the choice criterion. In general, we use the notation $u(z)$ to describe the utility of outcome z, and the notation $p(\theta)$ to describe the probability of the parameter(s) θ. The expected utility $\mathcal{U}(a)$ of action a is then

$$\mathcal{U}(a) = \int_{\mathcal{Z}} \int_{\Theta} u(z) p_a(z|\theta) p(\theta) d\theta dz. \qquad (2.3)$$

The average is taken over both unknown health outcomes and unknown parameter values. An optimal action, sometimes called 'Bayes' action, a_B, is one that maximizes the expected utility, that is,

$$a_B = \arg\max \, \mathcal{U}(a). \qquad (2.4)$$

Both outcomes and parameter values are unknown, both are assigned probability distributions, and both are integrated out in computing the expected utility $\mathcal{U}$. The main distinction between the two concepts is that the probability distribution of outcomes depends on the action taken, while that of the parameter values does not. How this distinction is drawn may change depending on the problem at hand. For example, whether a patient has cancer can be a parameter in a decision model to assess a strategy for early detection of occult disease, and can be an outcome in a decision model to assess a strategy for prevention of future disease.

Careful specification of the functions p_a, u and $p(\theta)$ is critical to decision-theoretic approaches. In practice, p_a and $p(\theta)$ will be conditional on all relevant data available at the time the decision is made, although this dependence is not made explicit in the notation. The discussion of Chapter 1 is relevant to the choice of $p(\theta)$. The choice of u will be discussed later in this Chapter, in the context of quantifying the utility of a major stroke. Before that, we digress briefly to look at the intellectual developments that led to the principle of expected utility, and present a theory that provides a connection between expected utility decision making and broader notions of rational behavior.

2.2.3 Historical background

Probability, utility, and the expected utility principle have long been an integral part of decision making. The idea that mathematical expectation should guide rational choice under uncertainty was formulated and discussed as early as the seventeenth century. During that time, an important problem was finding the fixed amount of money that would represent fair exchange for an uncertain payoff, as when purchasing health insurance. Initially, the prevailing belief was that the fair fixed amount would be the expected payoff. The St. Petersburg game (Jorland, 1987) revealed an example where the expected payoff is infinite but where, in the words of Daniel Bernoulli, 'no reasonable man would be willing to pay 20 ducats as equivalent'. Bernoulli's view was that 'in their theory, mathematicians evaluate money in proportion to its quantity while, in practice, people with common sense evaluate money in proportion to the utility they can obtain from it' (Bernoulli, 1738). Using this notion of utility, Bernoulli offered an alternative: he suggested that the fair sum to pay for a game of chance is the *moral expectation*, based on computing the expected value of a logarithmic transformation of the resulting wealth. His approach is an example of expected utility.

The use of probability to quantify a DM's knowledge about uncertain events originates from the work of Ramsey (1931) and de Finetti (1937) on subjective probability. The subjectivist approach provides a coherent foundation for probability statements about a broad typology of events, and can be integrated in a natural way into individual decision making. As we have seen in Chapter 1, a strength of this view is that a single measure of uncertainty is used for both population variation and imperfect knowledge about parameters. This provides a simple, general, and self-consistent formal mechanism for modeling the effects of accruing experimental evidence on knowledge.

Justifications for using the principle of maximization of expected utility as a general guide to rational decision making have also been the subject of extensive investigation. A fruitful approach has been to establish a correspondence between this principle and a set of basic axioms. These axioms are expressed in terms of preferences among actions, and are satisfied if and only if one's behavior is consistent with expected utility. This correspondence provides a foundation for and additional insight into expected utility decision making. The critical contribution in this regard is that of von Neumann and Morgenstern (1944), who developed the first and fundamental axiom system. Savage (1954) used the subjectivist approach, together with some of the technical advances provided by von Neumann and Morgenstern and by Wald, to develop a general and powerful theory of individual decision making under uncertainty, which provides the logical foundation of current Bayesian decision theory. A sketch of what would become Savage's development is outlined by Ramsey (1931). Here, we review a related but simpler formulation proposed by Anscombe and Aumann (1963). This formulation makes explicit the important concept of parameter-dependent utility, and avoids some of the more complex technical development in Savage. Critical discussions and analytical details can be found in Kreps (1988), whose formulation is followed here, and Schervish (1995).

2.2.4 Axiomatic foundations

To avoid complex technical developments, we assume that Θ and $\mathcal{Z}$ are finite lists of parameter values and outcomes; we denote by k the number of events in Θ. Every action is characterized by a known probability distribution on $p_a(z|\theta)$. If we take any two actions a and a' from $\mathcal{A}$, and a weight $\alpha \in [0, 1]$, we can define the composite action $\alpha a + (1 - \alpha)a'$, as the action characterized by the probability distribution

$$\alpha p_a(z|\theta) + (1 - \alpha)p_{a'}(z|\theta).$$

Every action that is a composite of two actions in $\mathcal{A}$ must also be in $\mathcal{A}$ for the theory to work. In a medical decision problem that applies to an entire cohort of patients, a composite decision could be interpreted as one in which a proportion α of the cohort receives treatment a and the rest receive treatment a'. For example, treatment could differ within a cohort depending on patients' insurance plan. An alternative interpretation is that α may reflect uncertainty at the level of individual patients. This occurs, for example, when a patient cannot be assumed with certainty to comply with the prescribed treatment. In that case α could be the (known) probability of compliance and a' the alternative course of action taken by the patient – for example, no treatment or a lower-dose treatment.

Axiomatization is in terms of statements like 'action a is preferred to action a''. This will be indicated by $a \succ a'$. Anscombe and Aumann (1963) showed that choosing actions consistently with the principle of expected utility is equivalent to holding preferences among actions that satisfy the following five axioms.

1. *Weak order axiom: The binary relation $\succ$ is a preference relation.* This condition means that comparisons between actions are *complete* (any two actions in $\mathcal{A}$ can be compared) and *transitive* (if $a \succ a'$ and $a' \succ a''$, then $a \succ a''$). This axiom is a critical building block in representing preferences for outcomes as points along a single dimension.

2. *Independence axiom: If $a \succ a'$ and $\alpha \in (0, 1]$ then, for every $a'' \in \mathcal{A}$,*

$$\alpha a + (1 - \alpha)a'' \succ \alpha a' + (1 - \alpha)a''.$$

This axiom requires that two composite lotteries should be compared solely based on the component that is different. It carries most of the weight in guaranteeing the 'expected' part in the expected utility principle.

3. *Archimedean axiom: If $a \succ a' \succ a''$, then there are $\alpha, \beta \in (0, 1)$ such that*

$$\alpha a + (1 - \alpha)a'' \succ a' \succ \beta a + (1 - \beta)a''.$$

To the DM, a is favorable, a'' is unfavorable, and a' is intermediate. If this axiom holds, the DM is always willing to find weights such that mixing in some of the unfavorable option with the favorable one will make the result worse than the intermediate option, and conversely for the unfavorable action. This axiom bars the DM from preferring an action to another so strongly that no reversal of preference is possible with any composite action that includes one of the two. This axiom carries most of the weight in representing the preference dimension as the real line.

4. *There exist a and a' in $\mathcal{A}$ such that $a' \succ a$.* The state θ is said to be *null* if the DM is indifferent between any two actions that differ only in what happens if θ occurs. This axiom is a structural condition requiring that not all states be null, which would make the problem moot.

5. *State independence axiom: For every $a \in \mathcal{A}$ and any two distributions q and q' on $\mathcal{Z}$, if*

$$\left(p_a(z|1), \ldots, p_a(z|\theta - 1), q, p_a(z|\theta + 1), \ldots, p_a(z|k)\right)$$
$$\succ \left(p_a(z|1), \ldots, p_a(z|\theta - 1), q', p_a(z|\theta + 1), \ldots, p_a(z|k)\right)$$

for some state θ, then, for all nonnull states θ',

$$\left(p_a(z|1), \ldots, p_a(z|\theta' - 1), q, p_a(z|\theta' + 1), \ldots, p_a(z|k)\right)$$
$$\succ \left(p_a(z|1), \ldots, p_a(z|\theta' - 1), q', p_a(z|\theta' + 1), \ldots, p_a(z|k)\right).$$

Before elaborating on the interpretation of axiom 5, let us consider a preliminary result. Using axioms 1–3, one can identify a function $f : \mathcal{A} \to \Re$ that represents the preference relation $\succ$, in the sense that an action is preferred to another whenever its f value is higher. The function $f(a)$ is linear in a. Using these two facts it can be shown (Kreps, 1988) that the first three axioms are necessary and sufficient for the existence of real-valued functions $v_1, \ldots, v_k$, such that

$$a \succ a' \iff \sum_\theta \sum_z v_\theta(z)p_a(z|\theta) > \sum_\theta \sum_z v_\theta(z)p_{a'}(z|\theta). \tag{2.5}$$

This result establishes a correspondence between preferences satisfying axioms 1–3 and preferences obtained by scoring actions using weighted averages. In this representation, the utility of a consequence and the probability of a health event are both captured by the term $v_\theta(z)$. This can be written as the product of a probability distribution on parameters and a utility function on outcomes, although this factorization is not unique.

Parameter-dependent utilities are common in medical problems, in which the uncertain parameter and the outcome are often intertwined. For example, women at high risk of having inherited a germline mutation of breast and ovarian cancer susceptibility genes *BRCA1* and *BRCA2* often consider preventive oophorectomy (the removal of the ovaries) to reduce the risk of ovarian cancer. After an oophorectomy, a woman cannot have children. In the absence of genetic testing, or following a negative, but potentially erroneous, test result, the critical unknown health-related event θ is whether the women is a carrier of a deleterious mutation of *BRCA1* or *BRCA2*. The truth about this event may bear on the utility of the outcomes of the oophorectomy. If the woman is a carrier her offspring will also be carriers with a probability of at least 50%. This consideration may affect how a woman values the ability of bearing a child.

Representation (2.5) is consistent with expected utility maximization, but does not provide a unique decomposition of preferences into probability of parameter values and utility of consequences. Such decomposition is achieved by the state independence axiom, which asks the DM to consider two comparisons: in the first the two actions are identical except that, in state θ, one has probability distribution p and the other has probability distribution q. In the second the two actions are again identical, except that now in state θ' one has probability distribution p and the other q. If the DM prefers the action with probability distribution p in the first comparison, this preference will have to hold in the second comparison as well. So the preference for p over q is independent of the state. This axiom guarantees separation of utility and probability. The main theorem about representation of preferences via utilities and probabilities is then: *Axioms 1–5 are necessary and sufficient for the existence of a nonconstant function* $u : \mathcal{Z} \to \Re$ *and a probability distribution* $p(\theta)$ *on* Θ *such that*

$$a \succ a' \quad \Longleftrightarrow \quad \sum_{\theta} p(\theta) \sum_{z} u(z) p_a(z|\theta) > \sum_{\theta} p(\theta) \sum_{z} u(z) p_{a'}(z|\theta).$$

Moreover, the probability distribution $p(\theta)$ *is unique, and* u *is unique up to a positive linear transformation.*

The monographs by Fishburn (1970; 1989) and Kreps (1988) provide insight, technical details and references on axiomatic expected utility theory. Because of the centrality of the expected utility paradigm in decision theory, axiom systems such as the one presented here have been investigated and criticized from both empirical and normative perspectives. Empirically, it is well documented that individuals sometimes violate the independence axiom (Allais, 1953; Kreps, 1988; Kahneman *et al.*, 1982; Shoemaker, 1982). Normative questions have also been raised about the weak ordering assumption (Seidenfeld, 1988; Seidenfeld *et al.*, 1995). Kadane *et al.* (1999) investigate the implications and limitations of various axiomatizations, and suggest possible generalizations.

2.3 Measuring the value of avoiding a major stroke

A stroke is potentially a severely disabling event. Prevention and treatment of stroke do aim at reducing mortality, but a critical target is also morbidity, in view of the difficulty of affecting mortality, and the substantial long-term impact of stroke-induced disabilities on both costs and quality of life of survivors (Matchar, 1998). Decision making regarding stroke-related medical intervention depends critically on understanding the value that individuals attach to avoiding the consequences of a stroke.

In this section we use this example to review and illustrate three approaches to the elicitation of individual utilities: the standard gamble method; the time trade-off method; and a more general two-step approach for measuring utility of health states over time. The goals are to illustrate the connection between the foundations and the elicitation, and to provide a practical context for the foundations. A wide variety of methods have been proposed for assigning value to health states. One category are rating scales, used to elicit utilities or quality directly. Another category are discrete-state health index models, which work by attaching fixed preference weights to observable health states. Comprehensive reviews include Nord (1992), Patrick and Erickson (1993), Kaplan (1995), Gold et al. (1996), and Lenert and Kaplan (2000).

2.3.1 Elicitation of personal utilities using the standard gamble approach

Let us fix the time horizon to the next ten years, and consider eliciting the utility of a major stroke over that time horizon. The so-called standard gamble approach to the elicitation of this utility stems straight from the direct comparisons between actions that were used to construct the axiomatization of utility theory. The first step is to identify two pivotal outcomes z_0 and z^0 that are easy to interpret and communicate and that can serve as the reference points in the utility scale. In assessing the utility of health states, 'death' is often chosen as z_0, while 'excellent health' or the 'status quo' are common choices for z^0. All utility functions that differ only by a change of scale lead to the same preferences, therefore we can arbitrarily set the utility of the two states z_0 and z^0 to $u(z_0) = 0$ and $u(z^0) = 1$ and express all others in terms of these two. This results in a convenient and interpretable utility scale.

In the stroke example, we may regard z_0 as 'death', z^0 as excellent health over the next ten years, and z as a major stroke and its sequela for a period of ten years, described in appropriate detail. We can then ask the patient to consider an action that results in excellent health with probability α and death with probability $1 - \alpha$. This action is called the standard gamble. The probability α refers to a hypothetical health event or parameter that may not need to be made explicit. The standard gamble has expected utility

$$\alpha u(z^0) + (1 - \alpha)u(z_0) = \alpha,$$

the same as the probability of the favorable outcome. Next we ask the patient to compare the standard gamble to an action that results in a major stroke for certain. The expected utility of this will be the same as $u(z)$, the utility of interest. The two alternatives are shown in Table 2.2. The utility for a major stroke over a ten-year horizon can be inferred by eliciting the value of α such

Table 2.2 The standard gamble methods elicits utility for the stroke state by finding the value of α such that the two actions a and a' are equivalent to the patient.

a	$\begin{cases} z^0 \text{ (excellent health)} \\ z_0 \text{ (death)} \end{cases}$	with probability α with probability $1 - \alpha$	
a'	z (major stroke)	with certainty	

that the patient is indifferent between the two actions. The expected utility of both actions would then be α, and therefore $u(z) = \alpha$ on a scale on which 0 is death and 1 is excellent health for ten years. Alternatively, a simple rescaling is to set $u(z^0) = 10$, the number of years of excellent health. This rescaling would produce a utility on a scale that is comparable to length of life. We will expand on this point in Section 2.3.2 when we consider the time trade-off method.

In this application, there may be patients who consider a major stroke to be worse than death, as we will see in Section 2.3.2. For those patients, there is no standard gamble for which the required indifference is reached. A simple modification is then required: the patient is asked to compare the standard gamble of Table 2.3 to death. The expected utilities are 0 for death and

$$\alpha u(z^0) + (1 - \alpha)u(z),$$

for the standard gamble, and are equal when $u(z) = -\alpha/(1-\alpha)$. One approach that avoids this complication is to choose the best and worst outcomes as pivotal. A drawback is that the best and worst outcomes are not necessarily as easily interpreted.

In general, the elicitation process can be described in terms of the following steps:

1. List outcomes.
2. Select all pivotal outcomes, one highly favorable (z^0) and one highly unfavorable (z_0), both easily described and interpreted.

Table 2.3 The standard gamble comparison for the case in which a major stroke is considered worse than death.

a	$\begin{cases} z^0 \text{ (excellent health)} \\ z \text{ (major stroke)} \end{cases}$	with probability α with probability $1-\alpha$	
a'	z_0 (death)	with certainty	

3. Assign arbitrary utilities of 1 to z^0 and 0 to z_0.

4. For all other outcomes z, determine whether it is worse than z_0, better than z^0, or intermediate.

5. If z is intermediate, elicit the probability α for which the patient is indifferent between z as a certainty and the standard gamble, and set $u(z) = \alpha$.

6. If not, proceed with a modified standard gamble.

If the patient's preferences are of the kind described by the axioms of Section 2.2.4, the existence of a unique value of α satisfying the indifference condition of step 5 above is implied by the Archimedean and independence properties of the DM's preferences. In practice, there can be serious difficulties in reliably measuring α. We consider some of these in Section 2.3.5.

2.3.2 Quality-adjusted life years and the time trade-off method

Outcomes of medical decisions are often chronic and best expressed as lengths of time spent in certain health states. For example, preventive measures may delay a stroke, increasing the length of time in the current state and possibly decreasing the length of time in the stroke state. In these circumstances, a popular alternative is based on the concept of the quality-adjusted life year (QALY), that is, the period of time in full health that a patient considers equivalent to a year in ill health. Put slightly differently, this means eliciting the number of years a patient is willing to trade off to avoid a certain chronic health state, such as a stroke. Compared to the standard gamble, this approach has the attraction of offering the DM options that do not involve probability, which is often difficult to conceptualize, or the possibility of immediate death. QALYs were pioneered by Fanshel and Bush (1970) and Torrance *et al.* (1972).

The Duke PORT group (Samsa *et al.*, 1999) carried out a survey to measure the QALYs for major stroke in a population at increased risk. Individuals are candidates for preventive intervention, which may range from watchful waiting, to highly invasive procedures, or surgery, such as carotid endarterectomy (Matchar, 1998). Subjects' utilities for a major stroke bear critically on the question of how aggressively prevention should be pursued. The patients population included individuals with previous transient ischemic attack or minor stroke, as well as others at increased risk of stroke because of conditions such as atrial fibrillation, hypertension, and valvular heart disease.

QALYs in the stroke state can be elicited using the time trade-off method. This method is based on considering an arbitrary period of time in a particular condition, and eliciting the amount of time in good health that is considered equivalent to it. For example, in the stroke application, interviewers described a major stroke as 'a stroke that leaves an arm, a leg, and one side of your body paralyzed, and leaves you unable to take care for yourself. Anyone who has a major stroke will stay in this state until death.' They then presented subjects with trade-offs like: 'Would you prefer living 10 more years after a major stroke or 8 more years in excellent health? In other words, would you give up 2 years of life after a major stroke in order to live 8 years in excellent health?' Time in excellent health was varied by the interviewer until a point of indifference was reached. In their protocol, the point of indifference was obtained by starting at 10 years and working down in increments of 1 year. Once the year was established, interviewers attempted to add 6 months, then finally 3 months. The resulting point of indifference was converted from a ten-year basis to a yearly basis by dividing years of life in excellent health by 10. For example, if the point of indifference is 7.25 years, then the QALYs for major stroke are 0.725.

The severity of a major stroke is such that some patients indicated that 10 years of major stroke was equivalent to 0 years in excellent health. Interviewers then followed up with the question: 'Given your feelings about the quality of life after a major stroke, which would you prefer: to live 10 years after a major stroke or to die from a major stroke the instant it occurred?'

Table 2.4 summarizes the results of the study. Patients were distributed across the full spectrum of possibilities, emphasizing the importance of individualizing the decision making process about stroke prevention. A substantial proportion considered a major stroke to be worse than death. For that proportion of the population, even aggressive or expensive stroke-prevention strategies may be appropriate.

Table 2.4 Utilities for a hypothetical major stroke in the Duke PORT study.

Utility range	Number of patients
Worse than death	543
0	10
0.025–0.10	59
0.125–0.20	22
0.225–0.30	37
0.325–0.40	57
0.425–0.50	88
0.525–0.60	74
0.625–0.70	93
0.725–0.80	85
0.825–0.90	59
0.925–1.0	59

2.3.3 Relationship between the standard gamble and the time trade-off method

QALYs are a practical and interpretable measure of quality of life in many medical decision problems, and have grown in popularity. For a health outcome z, they define a score $q(z)$ between 0 and 1 that can be multiplied by the time spent in a health state to model the consequence of health interventions with long-term effects. For example, imagine that surgical alternative a leads to perioperative mortality (immediate death) with probability 0.01 and to life with moderate disability for an exponentially distributed amount of time with mean 5 years. On the other hand, alternative a' leads to immediate death with probability 0.08 and to life with the same moderate disability for an exponentially distributed amount of time with mean 10 years. A single elicitation of QALYs for the moderate disability state can be sufficient to determine a choice between a and a'. If the moderate disability in question is valued, say, at 0.93 QALYs, the expected QALYs for a are

$$0 \times 0.01 + 0.93 \times 5 \times 0.99 = 4.6035,$$

while the expected QALYs for a' are

$$0 \times 0.08 + 0.93 \times 10 \times 0.92 = 8.556.$$

This procedure is similar to computing an expected utility, but are QALYs utilities in the sense of Section 2.2.4? Weinstein *et al.* (1980) identified conditions under which QALYs are consistent with expected utility theory. First, utility over health states must be such that there is independence between length of life and quality of life. This means that trade-offs between different lengths of life are not affected by the quality of life experienced by the individual. Similarly, trade-offs among levels of morbidity are not affected by the length of life that the individual will live. Second, utility over health states must display proportional trade-offs. This means the following. Suppose a person is indifferent between y years of life in health state S, say excellent health, and y' years of life in a worse health state S', say major stroke. Then he or she must also be indifferent between spending qy years of life in S and qy' years of life S. Finally, the individual must be risk-neutral with respect to years of life. A risk-neutral individual is one who is indifferent between an uncertain gain and its expected value with certainty. A *risk-averse* individual prefers the expected value with certainty. In terms of health-related decisions, a risk-neutral DM values the next year as much as any of the following years.

Table 2.5 illustrates the correspondence between the standard gamble and time trade-off methods. The standard gamble is based on the comparison of a and a', and leads, by construction, to utilities. The time trade-off is based on the comparison of a' and a''. The two comparisons give the same utility if $\alpha = q$. In that case alternative a'' gives, with certainty, the expected value of the length of time from alternative a. If this holds no matter what the

Table 2.5 Relationship between the standard gamble and time trade-off methods. The standard gamble is based on the comparison of a and a', while the time trade-off is based on the comparison of a' and a''.

a	$\begin{cases} z^0 \text{ (excellent health for 10 years)} \\ z_0 \text{ (death)} \end{cases}$		with probability $\quad \alpha$ with probability $1 - \alpha$
a'	z' (major stroke for 10 years)		with certainty
a''	z'' (excellent health for $q \times 10$ years)	with certainty	

horizon, as is the case when the DM is risk-neutral, then the two methods are equivalent and q can be interpreted as a utility.

Risk neutrality is uncommon in real situations. For example, Redelmeier and Heller (1993) asked participants in a study to consider three temporary events (colostomy, blindness, depression) that were hypothesized to occur at five different times in the future (one day, six months, one year, five years, and ten years) and measured their utility. Subjects were not risk-neutral (of all the discount rates, 62.1% were zero, corresponding to risk neutrality, 10.0% were less than zero, and 15.7% were greater than 0.10). Discount rates were not constant over time (mean discount rates for earlier time intervals tended to be larger than those for later ones) and also differed by type of health event.

If the time in health state z is a random variable y, application of QALYs as described in this section is based on computing the expectation of the quantity $q(z)y$. Pliskin *et al.* (1980) proposed an extension of this approach to adjust for potential risk aversion, by computing the expectation of $\left(q(z)y\right)^{\rho}$, where ρ is a risk aversion parameter. They characterize the type of personal preferences that fit this profile.

Even when an individual is very close to being risk-neutral, the time trade-off and standard gamble could lead to empirically different results, as they may capture different aspects of utilities, such as desirability of states, time preferences, and risk attitude, to different degrees. Nord (1992) and Kaplan (1995) review empirical evidence on the correlation of alternative elicitation methods, and conclude that statements about validity and correlation can only be made with respect to specific implementations rather than general approaches.

2.3.4   *Combining time trade-off and standard gamble to elicit utilities of risk-averse patients*

In our discussion so far we have encountered different sources of variation in individual attitudes toward a period of time in the major stroke state. First, the burden of a major stroke may vary from patient to patient, as shown in Table 2.4. Second, some patients are risk-averse, that is, they place greater value on nearer than on distant years. Finally, patients may be willing to trade off a certain proportion of their lives to avoid a major stroke if they had a normal life expectancy, but a smaller proportion if they had a shorter life expectancy. McNeil *et al.* (1981) described a general procedure to elicit utilities for length of life in ill health, which can account for all three of these sources of variation.

Their method proceeds in two steps. In the first step the interviewer will elicit the patient's preferences between living a shorter time with excellent health and a longer time with a major stroke, using the time trade-off method of Section 2.3.2. In the second step the interviewer will elicit the utility of different lengths of life in good health, using the standard gamble strategy as in Section 2.3.1. This provides an indirect mechanism for finding the patient's utilities for additional years with a major stroke, by first converting different lengths of life with a major stroke into equivalent lengths of life in excellent health, and then converting the result into utilities.

Table 2.6 illustrates this approach in a hypothetical example. Utilities refer to an individual patient, for a 20-year horizon. The left-hand and middle columns report the results of five assessments of the time trade-off between major stroke and excellent health. The first year in major stroke is valued as much as the first year in excellent health, but in subsequent years, the DM is willing to trade off some length of life to avoid the stroke related disability. Next the right-hand and middle columns represent five assessments of the standard gamble for the same times in excellent health that emerged from the time trade-off exercise. This utility function shows risk aversion; it is approximately the square root of the ratio between the number of years and the horizon.

One attraction of this approach is that it separates consideration of length from quality of life. While more demanding than the time trade-off , it reduces

Table 2.6 Two-step elicitation of patient's utilities for various lengths of life in major stroke. For example, the utility for 7 years of major stroke consequences is obtained by converting those 7 years into equivalent years of excellent health, and then converting the result, in this case 10 years, into utility, in this case 0.7

Years of major stroke	Equivalent years of excellent health	Utility
0	0	0.00
1	1	0.22
4	5	0.50
7	10	0.70
12	20	1.00

the number of questions that are necessary compared to a full gamble-based elicitation, and makes the task more manageable for the DM. To work well, it requires that the DM's risk aversion is relatively constant over health states.

2.3.5 Using individual utilities in medical decisions

Utility assessment in health care is complex. Difficulties have been documented extensively from psychological, medical, and economic viewpoints. In many situations, important components of the problem at hand may not be reliably quantified. These include religious or ethical beliefs, attitudes to coping with disease and dying, prejudice in favor of or against medicine or the medical profession, or the desire to know one's genotype or health status irrespective of the clinical use that may be made of this information. These factors can lead to preferences that are more complex than can be described within the axiomatization of expected utility theory, and even when that is not the case they can crate a challenge for the measurement of utilities. In addition there are a number of well-known cognitive difficulties in elicitation. Inconsistencies may arise depending on how the question is formulated, or on the description of the scenario for the outcomes. For example, patients tend to express different preferences when outcomes are presented as beneficial. Another problem is that, even though patients may have an intuitive understanding of the concept of probability, they usually have difficulties with the meaning of very low or very high probabilities. For additional discussion, see Patrick and Erickson (1993) and Kaplan (1995).

These limitations mitigate the enthusiasm for taking patient utilities at face value in decision modeling. Utility assessments, and the expected utility evaluations that ensue from them, are therefore used at various levels of stringency depending on the DM, the context, and the quality of the data sources.

One way in which individual utilities can be used directly is by interactive elicitation of preferences in counseling situations. A decision model could receive as input individual preferences in the form of an interview, and produce expected utility-based recommendations. For example, Scott *et al.* (1997) developed a computer architecture, called SecondOpinion, designed for automated web-based patient decision support:

> 'SecondOpinion custom tailors the discussion of therapy options for patients by eliciting their preferences for relevant health states via an interactive WWW interface and then integrating those results in a decision model. The SecondOpinion architecture uses a Finite State Machine representation to track the course of a patient's consultation and to choose the next action to take. The consultation has five distinct types of interactions: explanation of health states, assessment of preferences, detection and correction of errors in preference elicitations, and feedback on the implications of preference'.

In clinical decision making, a common approach is also to use small patient surveys, or clinical expertise, to support the choice of alternative quality of life scenarios that well represent typical patients. A decision model is then solved in each of the scenarios. Model users can informally categorize themselves or patients as being close to one of the proposed scenarios. An example of this approach is the decision analysis of axillary lymph node dissection in Chapter 5.

Often utility is used as a convenient way of combining multiple health outcomes on a unified scale. For example, tamoxifen for prevention of breast cancer also affects the risk of stroke and other adverse outcomes. Gail *et al.* (1999) used utility-like weighing schemes to develop a simple index reflecting the risks and benefits of chemoprevention using tamoxifen. They used the index to identify subgroups of women for whom chemoprevention may be beneficial. Another example is the use of average utilities from samples of patients in societal decisions. While ignoring possibly important individual variability, these provide a first-order approximation of the appropriate scale for weighing the costs and benefits of societal decisions.

2.4 Decision making in health care

In our discussion in Chapter 1 we examined how to develop probability distributions for describing uncertainty about future health events, and about scientific knowledge relevant to a decision. Earlier in this chapter, we also examined how to use utility theory to formally assign value to the health outcomes resulting from one's actions. A formal way of integrating these two components into a quantitative scheme for decision making is to perform expected utility calculations and use them to make medical decisions. In this section we discuss the role that this formal mechanism can play in practical decision making in health care, and we consider in detail the type of expected utility calculations that are involved in carrying it out.

2.4.1 Evidence–based medicine and synthesis of evidence

In medical problems, the increasing volume, complexity, and quality of medical information, the fragmentation of scientific investigations, the emphasis on hypothesis testing and risk factor identification, and the rarity of thorough reporting of uncertainties, make it hard to master and integrate the body of information that needs to be brought to bear in reaching good decisions. In parallel, evidence-based medicine, that is, the emphasis on explicit use of current best evidence in making decisions about the care of individual patients (Evidence-Based Medicine Working Group, 1992; Sackett *et al.*, 1996), is encouraging clinicians to be aware of the strength of the evidence in support of their clinical practice. As a result, medical decision making is evolving towards an evidence-and model-based scientific discipline.

Clinical guidelines are playing an increasing role in both medical care and clinicians' education, and models are becoming a critical components of guidelines.

What may constitute current best evidence for a particular decision problem depends, of course, on the problem itself and is generally complex to assess. It seems clear, however, that evidence-based medicine requires the integration of individual clinical expertise with the best available external clinical evidence from systematic research, and that the latter comes from diverse sources, including randomized clinical trials, cohort and case–control studies, surveillance data, administrative data, and patient preference data. In this scenario, a concern is how evidence and expert judgment should be interpreted, integrated, communicated, and effectively linked to medical practice. In this regard, Matchar and Samsa (1999) remark that:

> Incorporating multiple data sources in a way that informs complex clinical decisions is a substantial analytical challenge. One approach to this challenge is to develop a simulation/decision model that explicitly represents the natural history of disease and the impact of treatments on that natural history. The model should be requisite – that is, sufficient in form to address the decision problem – but not overly complex. Such a model can be of value because it (1) allows a variety of viewpoints to be considered, (2) incorporates the best scientific evidence, and (3) permits sensitivity analyses to evaluate the impact of alternative clinical scenarios and uncertainty in model inputs.

A formal decision model can, via its probabilistic component, synthesize current best evidence and clinical judgment about the treatment or prevention of a medical condition, and can, via its utility component, link this evidence to clinical decisions. While DMs are understandably reluctant to delegate the decision making process to model-based formalisms, they are increasingly appreciating the role of quantitative modeling in linking evidence to decisions, and are relying on structured approaches to make more informed decisions.

While expected utility theory forms the backbone of Bayesian decision theory as well as much of its mathematical foundations, the role of Bayesian decision models in medicine should go beyond that of simply determining a single decision maximizing expected utility. The process of developing a model along Bayesian lines, documenting the expert- and evidence-based input that contributed to the modeling, translating effect sizes and tests of hypotheses into outcome predictions, quantifying the uncertainties that emerge from the model, and exploring the entire utility surface as it varies over multiple dimensions depending on the many factors involved, all are integral parts of Bayesian modeling and provide powerful quantitative backing to decision making. In this broader sense, Bayesian decision modeling can be used to inform and guide decision making using formal structures that differ in extent and scope depending on the problem and the agents involved, rather than necessarily subsuming decision making in its entirety.

Reliable decision models are hard to develop, and it is important to document them, keep them up to date, make them widely available, have them validated by independent parties, and reuse them in several related decision problems (Matchar et al., 1997). The idea of a multiuser model stretches the limits of standard Bayesian decision theory, in which the DM and the model builder are one and the same, and in which it is often possible to say to a good approximation what is the purpose of the model. Decision making can still drive model development, but a model developer is working to enable informed decision making by as many different agents as possible. There is not usually a single purpose.

Weinstein's (1989) seminal work on cardiovascular disease modeling demonstrated that it is useful to build comprehensive decision support models that can serve as the basis for multiple decision problems and multiple DMs. He remarks that:

> Policy models are intended to guide the choices of persons and organizations that affect the aggregate allocations of resources to health care problems. Although it is difficult to identify any single policymaker in the United States who can alter the aggregate effect of the millions (or billions) of individual clinical decisions, there are many potential users of policy models: payers, providers, state and local health departments, the National Institutes of Health, professional organizations, hospitals and producers of medical devices, among others.

In addition to directly furthering evidence-based decision making, decision models can play an important role in the scientific community. A model can serve as the organizing focus of a research field, by helping identify the most critical gaps in knowledge and the most relevant controversies. It can formally serve to project results of planned trials, helping the planning and prioritization of research efforts. It can assist the exploration of hypothetical scenarios that have not been considered directly in randomized trials, such as novel diagnostic technologies, proposed legislation, or clinical recommendations. Examples of decision models that share some of these goals are the Coronary Heart Disease Policy Model (Weinstein et al., 1987) and the Stroke Prevention Policy Model (Matchar et al., 1997). Developing, validating, and maintaining reliable comprehensive models is a challenge, but the benefits of such models are likely to outweigh the costs of continuing to rely on less systematic and quantitative methods.

2.4.2   Elements of a decision model

From a statistical viewpoint, a decision model provides a joint predictive distribution of outcomes – utilities, costs, survival and so on – under various strategies, conditional on a patient's predisposing factors. From this joint predictive distribution one can then derive the expectations that are typically used in practice for comparing interventions, including all quantities that are necessary for individual decision making or for a cost-effectiveness analysis. Depending on whether the results are to provide guidance to individual or policy decision making, patient-specific characteristics can be conditioned upon or they can be integrated out based on their marginal distributions in a population of interest.

Predictive distributions can seldom be estimated in the required level of detail based directly on a single study, say a randomized trial data. More often, one needs to integrate evidence from several sources. While this requires assumptions that are sometimes difficult to test, it provides the opportunity to use each source for the purpose to which it is best suited. For example, population-based epidemiological data are typically more reliable than randomized trials data for modeling aspects of the diseases natural history, such as prevalence or effects or risk factors. But clinical trials are more appropriate for modeling intervention effects.

To help make the presentation of these concepts more concrete, consider again the case of the μSPPM. In our discussion in Section 1.5.1 we presented a Markov process, depicted in Figure 1.11, for describing the history of stroke-related events in a cohort at risk, under standard care. This process can provide the basis for a decision model for comparing alternative interventions. To illustrate, consider a hypothetical surgical procedure to be performed on TIA patients to reduce the risk of stroke. We will assume that the procedure has a perioperative mortality, and that only a fraction of patients survive the operation. Those who do survive experience a reduction in the risk of stroke, with stenosis patients benefiting to a greater extent. A decision model can help address the trade-off between perioperative mortality and decreased stroke risk, and determine the best course of action for each group of patients. We will present details of this example in Section 2.5.2, after giving a general view of the necessary expected utility calculations.

For ease of exposition, we distinguish the following elements of a decision model:

- health states and transitions;
- effects of covariates;
- effects of diagnostic tests;
- intervention effects;
- health outcomes and costs.

We comment briefly on each of these elements in the context of the μSPPM. Transition and duration distributions assume that the decision model considers explicitly the progress of a patient's health over time. This is not always necessary, but the general ideas of this discussion also apply to models with a short time horizon or no dynamic component at all.

The core of a decision model is a representation of the progress of the disease, in the absence of clinical intervention, or under standard care. This begins with an exhaustive and mutually exclusive classification of a patient's health states at a given time or age. Initially, specification of possible health states is based on informal incorporation of published results and expert opinion. Subsequently, the list and configuration of states may change as the result of statistical estimation of other parts of the model. In particular, distinguishing between clinically different health states is important only if there also is a difference in terms of the patient's prognosis, utilities, costs, or other outcomes.

The health states provide a static representation of an individual's health. A model for the progress of the disease over time requires a description of the transitions between states. In the μSPPM example of Figure 1.11, these are represented by the arrows, and mimick a typical pattern for a chronic disease. For example, patients do not revert to the HEALTH state after entering the TIA state. It is useful to separate these individuals from the rest because they might be the target of specific stroke prevention interventions, they may have higher medical bills, and so forth.

The time at which transitions occur is usually modeled in terms of the patient's age, rather than chronological time. This in turn may be discrete or continuous. A convenient way of specifying the probabilistic structure of the model is to assign a duration distribution to each of the states, and a destination distribution upon exiting the state, as discussed in Section 1.5.2. Estimation of the duration distribution is typically based on epidemiologic studies. From a statistical viewpoint, there are several challenges. First, this is the stage at which it is necessary to model patient- and provider-specific features. This is typically done by means of generalized linear models, or proportional hazard models, with covariates. Second, when there is a high number of duration distributions and covariates, estimation must often resort to information derived from different data bases or studies. In some cases, little is lost by focusing on one good study or data set properly addressing each of the questions of interest. Often, however, the limitations of individual data sets make it necessary to formally combine information from different sources. Useful tools for addressing both problems can be developed within the framework of Bayesian hierarchical models, discussed further in Chapter 4.

The model components described so far can be used to evaluate the implications of no intervention, or standard care. To determine the implications of other available clinical strategies, an effective option is to modify the incidence rates in the natural history model, using intervention-specific risk ratios or other adjustments. In the μSPPM example we will use this approach to study an intervention that reduces the risk of stroke in

TIA patients. Intervention effects are obtained by reviewing and analytically synthesizing published research and other information on clinical strategies. Ideally, the evaluation of risk ratios should be based on randomized clinical trials. Interventions could potentially be very complex. For example, only certain subsets of patients may receive treatment, depending on their characteristics, previous medical history, previous interventions and so on. Combining information on treatment effects from multiple studies is often done via meta-analysis, discussed in Chapter 4.

Diagnostic tests share some of the features of covariates, because they reveal patient characteristics that can affect their transition probabilities, as well as features of intervention, because they may imply morbidity and/or a direct clinical effect. For example, when performing a colonoscopy for the early detection and prevention of colorectal cancer, there is the possibility of having both complications and a preventive effect if precancerous polyps are found and removed. Modeling of diagnostic testing is complicated by the interplay between the direct effect of the test and the value of the information provided by the results. One illustration is given in Chapter 5.

The consequences of clinical interventions are usually evaluated based on a number of different outcomes. Formally, the problem is probably best captured by a multiway outcome function, whose dimensions may include survival times, patient individual utilities, financial costs and so on. In the context of a stochastic compartment model such as the μSPPM, costs and individual utilities can be incorporated by attaching a value to each health state and each transition. These represent respectively the cost of remaining in the given health state for one unit of time, and the cost of a change in health state. Together with the transition distribution, these define a cost process over time. Cost estimation is typically based on administrative data sets.

2.4.3 Variation and uncertainty in decision modeling

In view of this discussion, the simple dichotomy of events and outcomes used so far in describing expected utility theory needs to be developed in greater detail, to better characterize the differences between individual and policy decision problems, and to better identify the role of statistical inference tools. First, it is useful to differentiate explicitly between patient variation within a cohort, and uncertainty in knowledge. Both are described by probabilities in a Bayesian analysis, and both are ultimately subjective in that they represent the DM's understanding of medical knowledge, but they concern different aspects of this knowledge.

Patient variation refers to differences among individuals in the population from which the individual is drawn, or to which the policy will be applied. Important sources of individual variation include:

- Variation in the risk factors that affect the probability of different health events. Example are family history in the genetic testing application of Section 1.3 and risk factors for cardiovascular disease, such as stenosis, in the μSPPM.
- Variation in outcomes, including financial outcomes, for a given set of risk factors and interventions. In the μSPPM, this would be described by the Markov process of Figure 1.11.
- Variation in utilities for a given health outcome, deriving from the different values individuals attach to the same health state, as illustrated by the major stroke example of Table 2.4.
- Variation in individual implementation of the interventions. Many health interventions that require individuals to change their behavior, are implemented to different degrees by different individuals. For example, compliance with drug schedules can be problematic for chronic patients.

Parameter and model uncertainty arise from limitations in the sources of information used in building and estimating the model. They can represent sampling variability in data sources, or, more broadly, limitations in the DM's knowledge about model features. Such uncertainty in medical knowledge can arise from several sources, including:

- Imperfect knowledge of the extent to which the data sources utilized to develop a model do apply to the population or patient under consideration;
- Imperfect knowledge of the appropriateness of model assumptions, such as the exclusion a risk factor from a model, or the Markov assumption;
- Imperfect knowledge of model parameters.

Generally, quantitative investigation of each of these components of uncertainty can make a critical contribution to decision making. Expected utility decision making integrates out both individual variation and imperfect knowledge. An alternative approach is to integrate out only the individual variation, and study the effect of imperfect knowledge on the conclusions using tools such sensitivity analysis. Additional discussion of this approach will be found in Section 3.4.1.

As in Section 2.2.1, we denote by z the future history of relevant health outcomes, and by x the vector of risk factors or covariates. Both x and z are potentially of very high dimension. For each action, the *outcome model* is a probability distribution of the outcomes for an individual with covariates x if action a is taken. Formally,

$$p_a(z|x,\theta), \tag{2.6}$$

where θ represents the unknown model parameters, as earlier. Variation in individual implementation of the interventions can also be incorporated directly in the outcome model (2.6). The variation of covariates in the population of interest is described by a population model,

$$p(x|\theta). \tag{2.7}$$

The vector θ comprises parameters of both the outcome and the population model. The two sets could overlap. The distinction between outcomes and covariates may not always be easy to draw, especially when covariates are time-varying. For the purpose of this discussion, covariates are such that they cannot be influenced by medical actions. For example, if the decision problem is early diagnosis of lung cancer, smoking can be a risk factor, while if the decision problem is evaluating lung cancer prevention interventions it may be part of the outcome.

Within a Bayesian analysis, knowledge of model parameters and of modeling assumptions is represented by posterior probability distributions. Formally, model specifications such as whether to include a covariate, or whether to use a gamma or Weibull distribution for describing the time to a certain event, can be considered part of the catch-all parameter vector θ. For example, inclusion of a covariate can be captured by a simple binary parameter. Technically, this poses computational and other challenges (Draper, 1995), but conceptually it allows us to use a unified approach for both model and parameter uncertainty. If we denote by "Data" the entire body of evidence brought to bear in developing the model, uncertainty about θ is captured by the distribution

$$p(\theta|\text{Data}). \tag{2.8}$$

These three distributions fully characterize the uncertainties that need to be modeled in the decision problem.

2.4.4 Expected utilities

We now consider in detail the expected utilities of alternative actions in both individual and policy decision making. Consider first the case of a single individual with utility u and covariates x. From this individual's viewpoint, the relevant expected utility of action a is:

$$\mathcal{U}(a) = \int_\Theta \int_{\mathcal{Z}} u(z) p_a(z|x,\theta) p(\theta|\text{Data}) dz d\theta. \tag{2.9}$$

This quantity depends on x as well as a; maximization leads to an optimal choice that is specific to the individual's values and risk factors. The case study of Chapter 5 will provide an example. Implicit in expression (2.9) is an assumption of conditional independence of future outcomes z and past data given the parameter vector θ. This is usually plausible, as θ often adequately summarizes prior data for the purpose of model (2.6). Because of this conditional independence assumption, we have that

$$\int_\Theta p_a(z|x,\theta) p(\theta|\text{Data}) \, d\theta = p_a(z|x,\text{Data}),$$

which is the posterior predictive distribution of Section 1.4.6. Expression (2.9) can also be rewritten as

$$\mathcal{U}(a) = \int_{\mathcal{Z}} u(z)p_a(z|x, \text{Data})\, dz. \tag{2.10}$$

Let us now consider a policy action a that affects an entire population of individuals, such as spraying a residential area with pesticides in the recent measure to prevent the spread of West Nile disease. If these individuals have different covariates, the distribution of covariates needs to be described by a population model like (2.7). If we assume that all individuals have the same utility u, and all are given equal weight in the decision, the relevant expected utility of action a is

$$\mathcal{U}(a) = \int_{\Theta} \int_{\mathcal{X}} \int_{\mathcal{Z}} u(z)p_a(z|x, \theta)p(\theta|\text{Data})dzdxd\theta. \tag{2.11}$$

This quantity does not depend on x, but only on the mixture of covariates in the population being studied.

The assumption of a common utility to all individuals is often violated. Ideally, in developing societal decisions, variation in patient utility should be integrated out. One approach is to describe, using samples of utility from patient surveys, a distribution of utilities for a given health outcome z, possibly as a function of covariates. If we call this function $p(u|z, x)$, and assume that utility varies in the range from 0 to 1, we can express the relevant expected utility of action a as

$$\mathcal{U}(a) = \int_{\Theta} \int_{0}^{1} \int_{\mathcal{X}} \int_{\mathcal{Z}} up(u|x, z)p_a(z|x, \theta)p(\theta|\text{Data})dzdxdud\theta. \tag{2.12}$$

This gives each individual the same weight and assumes that utilities can be standardized and directly compared across individuals. This view can be traced back to utilitarianism. For example, in 1789, in *An Introduction to the Principles of Morals and Legislation*, Jeremy Bentham (see Bentham, 1876) wrote:

> The interest of the community is one of the most general expressions that can occur in the phraseology of morals: no wonder that the meaning of it is often lost. When it has meaning, it is this. The community is a fictitious body, composed of the individual persons who are considered as constituting as it were its members. The interest of the community then is, what? – the sum of the interests of the several members who compose it.

Aggregation of individual preferences has been challenged, but is still commonly applied. Kaplan (1995) offers a discussion of the role of variability of individual preferences in societal decision, and reviews the debate on the aggregation of individual utilities in determining societal choices for health care.

2.4.5 Plug-in approximation to expected utility

The Bayesian approach to prediction and expected utility calculations, based on averaging over unknown values of the parameters, can be contrasted with the commonly adopted *ad hoc* approach of approximating expected utility calculations by replacing θ with an estimate $\hat{\theta}$, Bayesian or classical, and ignoring the integration with respect to θ. This approach is sometimes called 'plug-in' approximation. Formally, in the individual decision case, the plug-in approximation of (2.9) is

$$\mathcal{U}(a) \approx \int_{\mathcal{Z}} u(z) p_a(z|x, \hat{\theta})\, dz.$$

The plug-in approximation is practical computationally, but it leads to the use of outcome distributions that are underdispersed, as no account is taken of uncertainty about the parameter estimates.

To illustrate the potentially serious flaws of the plug-in approximation in evaluating expected utilities, consider again the genetic effect modifier example of Section 2.2.1. Table 2.7 gives hypothetical expected recovery times for each combination of parameter and treatment: under treatment A recovery for patients with responsive genotype is fast (3 months on average), while for patients with unresponsive genotype it is slow (24 months on average); under treatment B, normal recovery is 6 months on average, irrespective of genotype. Suppose that, based on a well-conducted study, we can estimate that for a certain patient $p(\theta = 1|\text{Data}) = 0.8$. The expected recovery times are $3 \times 0.8 + 24 \times 0.2 = 7.2$ months for treatment A, and 6 months for treatment B. A utility function that rewards short recovery would lead to treatment B being prescribed. On the other hand, if we were to replace θ with a best estimate, we could set $\hat{\theta} = 1$, the most likely value. Then the plug-in approximation of the expected recovery times would be 3 months for treatment A and 6 for treatment B, and A would be prescribed.

In this example, the discrete support of the parameter makes the disadvantage of the plug-in estimate easy to see. In continuous parameter spaces, the plug-in approximation may be less problematic in practice. In general, the plug-in estimate produces reliable answers when the integrand in (2.9) is approximately linear as a function of θ, or when the posterior

Table 2.7 Expected recovery times in months, for given parameters and treatment choice in the genetic modifier example.

	Responsive genotype ($\theta = 1$)	Unresponsive genotype ($\theta = 0$)
Treatment A	$E_A(z|\theta = 1) = 3$	$E_A(z|\theta = 0) = 24$
Treatment B	$E_B(z|\theta = 1) = 6$	$E_B(z|\theta = 0) = 6$

distribution of θ shows minimal variability. Additional discussion of this issue is given in Section 3.3.2.

2.5 Cost-effectiveness analyses in the μSPPM

2.5.1 From decision modeling to cost-effectiveness

Our discussion so far has considered outcomes in the abstract. Utility theory can in principle handle outcomes that are quite arbitrary and of high dimensional. For example, we could apply the standard gamble approach to the elicitation of utility for future histories that include both medical and financial details. In practice, however, it is often useful to separately evaluate the monetary implication of a medical intervention from its health-related implications. The contrast between these two sets of effects of an intervention is the basis of cost-effectiveness analysis (CEA), a widely used tool in medical decision making. Broadly, the goals of CEA are to establish whether an intervention provides a benefit at an acceptable cost (Doubilet et al., 1986), primarily with a view toward deciding how to allocate scarce health care resources (Sloan 1995). A formal decision-theoretic formulation of cost-effectiveness, using the dollar per QALY framework, is discussed by Pliskin et al. (1980).

In terms of the setting that we have developed so far, CEA can be carried out as follows. We consider CEA conditional on a covariate profile x, and costs. Generalization to a population requires additional integration with respect to x, as was done in Section 2.4.4. For a given outcome z, define $e(z)$ to be the efficacy of the intervention, measured in life years, QALYs or other patient utilities, but not including financial considerations. Also, define $c_a(z)$ to be the cost associated with outcome z when intervention a is used. Here we assume that the effectiveness depends on the action taken only through the health outcomes, while cost can be affected by the action both directly (e.g. cost of a procedure) and indirectly (prolonged life, reduced disability). Defining an appropriate strategy for evaluating costs in CEA is complex and can be controversial. A comprehensive discussion is provided by Gold et al. (1996). The basis of CEA are the expectations

$$\mathcal{E}(a) = \int_\Theta \int_\mathcal{Z} e(z) p_a(z|x,\theta) p(\theta|\text{Data}) dz d\theta, \qquad (2.13)$$

$$\mathcal{C}(a) = \int_\Theta \int_\mathcal{Z} c_a(z) p_a(z|x,\theta) p(\theta|\text{Data}) dz d\theta. \qquad (2.14)$$

The average cost-effectiveness ratio (ACER) of action a is simply the ratio $\mathcal{C}(a)/\mathcal{E}(a)$.

Cost-effectiveness is often measured relative to alternative strategies, such as the status quo or standard care. The marginal cost-effectiveness ratio (MCER) for intervention a', relative to intervention a, is defined as the difference in average cost divided by the difference in average effectiveness. For a patient with covariates x, this is

$$\text{MCER} = \frac{\mathcal{C}(a') - \mathcal{C}(a)}{\mathcal{E}(a') - \mathcal{E}(a)}. \tag{2.15}$$

A controversy in CEA concerns discounting, that is, valuing outcomes that are in the near future more than outcomes that are in the distant future. The issue is not only what discount rate to use but also whether both health outcomes and costs should be discounted or only the latter. A common approach is to use common discount rates for both effectiveness and costs. We refer to Gold *et al.* (1996) for further discussion. None of our examples includes discounting, as the focus is on the statistical aspects.

General reviews of statistical issues arising in CEA are provided by Mullahi and Manning (1996), Manning *et al.* (1996), and Siegel *et al.* (1996).

Another delicate issue in CEAs concerns the proper way to aggregate individual values. A simple example will illustrate the complexity of making decisions that affect an entire population, when different subgroups have different utilities and costs. Consider a hypothetical decision in which there are three options and two types of individuals, A and B, equally represented in the population. QALYs and costs for each type of individual and for the population average are shown in Table 2.8.

For type A individuals action 3 is the most cost-effective, while for type B individuals action 1 is the most cost-effective. If we average QALYs over the population, the most cost-effective option is action 2, which is not the best choice for any of the subgroups, and in this case is even dominated (i.e. beaten in terms of both costs and QALYs) in both population subgroups. Ideally, we would like to be able to implement action 1 for type B and action 3 for type A individuals, for example by identifying appropriate individual risk factors.

Table 2.8 Expected QALYs and costs in a hypothetical population with two patient types.

	Expected QALYs			COST
	Type A	Type B	Average	
Action 1	5	9	7	2.03
Action 2	8	8	8	2.25
Action 3	9	5	7	2.03

2.5.2 The μSPPM example

We now illustrate these general concepts using the μSPPM. We are interested in the MCER for surgical procedure a', compared to standard care a. We will assume that the procedure has a perioperative mortality, and that only a proportion ρ of patients survive the operation. Those who survive experience a reduction in the risk of stroke, with stenosis patients benefiting to a greater extent. Specifically, for patients with no stenosis ($x = 0$) the reduced risk is $\kappa^0 \lambda_{TS}^0$, while for patients with stenosis ($x = 1$) the reduced risk is $\kappa^1 \lambda_{TS}^1$. Here we have a trade-off between perioperative mortality and decreased stroke risk, the net effect of which needs to be weighed against the financial implications of the procedure. As effects differ, the analysis needs to be carried out separately for patients with and without stenosis.

We begin by reviewing the hypothetical data used to estimate the model. Table 2.9 summarizes the sufficient statistics used to develop the posterior distributions of the λs, using expression (1.28). In addition, we assume that a published clinical trial indicates that, as a result of the intervention, the hazard of making a transition from the TIA state to the STROKE state is estimated to be reduced by 23% in stenosis patients and by 5% in other patients. We assume that such a reduction is independent of the amount of time spent in the TIA state. The risk multipliers κ^0 and κ^1, characterizing the efficacy of the intervention, are 0.95 and 0.77, respectively. Based on the uncertainty assessment reported in the original study, we hypothesize normal distributions for κ^0 and κ^1, with means 0.95 and 0.77, and standard deviations 0.05 and 0.03. Perioperative mortality is observed in 3% of patients in the published study, so the parameter ρ, the probability of surviving the procedure, is modeled as a Beta (97,3) random variable.

As a measure of effectiveness we will take life years. We also consider hypothetical costs $c_T = 0.22$ and $c_S = 4.4$ for each unit of time spent in the TIA state and in the STROKE state, respectively. An additional cost $c_B = 2.3$ will be incurred if the intervention takes place. These are, say, in thousands of dollars. For simplicity, costs are assumed to be known and not

Table 2.9 Hypothetical sufficient statistics for the μSPPM example. n_{HT} is the number of transitions from HEALTH to TIA in the sample, and t_H the total time at risk in the HEALTH state. Statistics for other transitions are defined similarly.

t_H^0	=	5000	t_H^1	=	5000
t_T^0	=	500	t_T^1	=	236
t_S^0	=	350	t_S^1	=	375
n_{HT}^0	=	514	n_{HT}^1	=	396
n_{HD}^0	=	231	n_{HD}^1	=	348
n_{TS}^0	=	463	n_{TS}^1	=	220
n_{TD}^0	=	51	n_{TD}^1	=	29

discounted. No cost is considered for the H state. Because the intervention does not affect time in H, the MCER is unaffected by this. Extensions in both direction are straightforward. In this example we can use results on first passage times in finite-state Markov processes (Bhat, 1984) to evaluate the expected life years and costs in closed form, conditional on the parameters. This gives a direct view of the impact of intervention on the quantities that are of interest in CEAs.

Under standard care, the expected life years for a patient with covariates x are

$$\frac{1}{\lambda_H^x} + \frac{\lambda_{HT}^x}{\lambda_H^x}\frac{1}{\lambda_T^x} + \frac{\lambda_{HT}^x}{\lambda_H^x}\frac{\lambda_{TS}^x}{\lambda_T^x}\frac{1}{\lambda_{SD}^x} + \frac{\lambda_{HS}^x}{\lambda_H^x}\frac{1}{\lambda_{SD}^x}. \qquad (2.16)$$

Recall that $\lambda_H^x = \lambda_{HT}^x + \lambda_{HS}^x + \lambda_{HD}^x$ and $\lambda_T^x = \lambda_{TS}^x + \lambda_{TD}^x$. Expression (2.16) corresponds to the inner expectation in (2.13). Each of the terms in (2.16) has a simple interpretation: $1/\lambda_H^x$ is the expected life years in H, $1/\lambda_T^x$ the expected life years in T, and $1/\lambda_{SD}^x$ the expected life years in S. Also, $\lambda_{HT}^x/\lambda_H^x$ is the probability of moving to T when leaving H, and other ratios have analogous meanings. So (2.16) is interpretable as a weighted average of times in a given state, weighted by the probability of reaching that state. The time in the stroke state appears twice because it is convenient to consider separately the case in which it is reach directly and that in which it is reached via a TIA.

Deriving a corresponding expression for patients undergoing surgery requires two modifications: the transition rate from T to S is now $\kappa^x\lambda_{TS}^x$. This is smaller, leading to a longer stay in T and a longer life expectancy. The probabilities of reaching the various states are affected by the perioperative mortality. If we assume that TIA patients receive the treatment not long after the TIA event, the perioperative mortality reduces the TIA cohort, and also shortens life expectancy. These considerations are captured by the following expression for the expected life years for a patient with covariates x receiving surgery if a TIA occurs:

$$\frac{1}{\lambda_H^x} + \rho\left[\frac{\lambda_{HT}^x}{\lambda_H^x(\kappa^x\lambda_{TS}^x + \lambda_{TD}^x)} + \frac{\lambda_{HT}^x\kappa^x\lambda_{TS}^x}{\lambda_H^x(\kappa^x\lambda_{TS}^x + \lambda_{TD}^x)\lambda_{SD}^x}\right] + \frac{\lambda_{HS}^x}{\lambda_H^x\lambda_{SD}^x}. \qquad (2.17)$$

This expression is conditional on the values of ρ κs and λs and is unknown.

Similarly, we can determine expressions for the costs, given the parameters. The expected cost for a patient with covariates x under standard care is

$$c_T\frac{\lambda_{HT}^x}{\lambda_H^x\lambda_T^x} + c_S\left[\frac{\lambda_{HT}^x\lambda_{TS}^x}{\lambda_H^x\lambda_T^x\lambda_{SD}^x} + \frac{\lambda_{HS}^x}{\lambda_H^x\lambda_{SD}^x}\right], \qquad (2.18)$$

while patients receiving surgery have expected cost

Table 2.10 Elements of cost-effectiveness evaluation of surgery relative to standard care for patients with and without stenosis.

Stenosis	a: usual care		a': surgery	
no	$\mathcal{E}(a)$ = 5.856	$\mathcal{E}(a')$	=	5.857
no	$\mathcal{C}(a)$ = 1.652	$\mathcal{C}(a')$	=	3.919
yes	$\mathcal{E}(a)$ = 5.827	$\mathcal{E}(a')$	=	5.874
yes	$\mathcal{C}(a)$ = 1.768	$\mathcal{C}(a')$	=	4.034

$$c_B + c_T\rho\frac{\lambda_{HT}^x}{\lambda_H^x(\kappa^x\lambda_{TS}^x + \lambda_{TD}^x)} + c_S\rho\frac{\lambda_{HT}^x\kappa^x\lambda_{TS}^x}{\lambda_H^x(\kappa^x\lambda_{TS}^x + \lambda_{TD}^x)\lambda_{SD}^x} + c_S\frac{\lambda_{HS}^x}{\lambda_H^x\lambda_{SD}^x}. \quad (2.19)$$

To compute the MCER for the surgical intervention relative to standard care, we need to compute the expectation of the four expressions above with respect to the distributions of the parameters. This expectation is conveniently done using Monte Carlo methods described in Section 3.3.2. We used 1 million draws. Table 2.10 summarizes the results. For patients with stenosis, the difference in effectiveness is 0.0473 years, while the difference in costs is 2.267 thousand dollars, leading to an MCER of 47.90 thousand dollars per year of life saved. For patients with no stenosis, the difference in effectiveness is only 0.00104 years, while the difference in costs is 2.267 thousand dollars, leading to an astronomical MCER of 2182 thousand dollars per year of life saved. The reason for this high figure is the small denominator. Small denominators in MCERs raise two important issues from a statistical standpoint. First, MCERs are very sensitive to variations in the denominator. We will elaborate on this point in Section 3.4.1, where we study the entire distribution of cost, effectiveness, and the MCER. Second, the plug-in approximation can be misleading in the presence of small denominators. For example, in this application, the plug-in approximation of the MCER for the stenosis cohort is 48.41, a result of acceptable quality. For the no-stenosis group, however, the plug-in approximation of effectiveness of surgery is smaller than that of standard care (the difference is -0.0000341). The plug-in approximation leads to the erroneous conclusion that surgery is both more expensive and less effective.

2.6 Statistical decision problems

2.6.1 Statistical decision theory

Bayesian statistical decision theory is concerned with the application of the decision approaches discussed earlier in this chapter, and in particular the expected utility principle, to choosing among actions that represent inferences, predictions, or conclusions to be drawn from experimental evidence. This perspective brings basic principles of rationality to bear in developing and evaluating statistical methodologies. More specifically, a statistical decision problem is one in which the utilities depend on some unknown parameter θ, and there is the possibility of observing the outcome y of an experiment with probability distribution $p(y|\theta)$. Observing y can give information about θ, and assist in decision making. Often, the utilities are set in abstract terms, to investigate the general properties of inferential procedures. This approach offers ways of making efficient use of the information provided by y in choosing an action, and ways of measuring the value of an experiment and deciding whether or not to perform it when it comes at a cost.

Historically, the notion that, when weighing evidence provided by data, one needs to take explicitly into account the consequences of the actions that are to be taken using the data goes back to the age of Enlightenment. Condorcet (1785), for example, reasoned that 'The probability that a convicted person is guilty should be to the probability that an acquitted person is innocent as the inconvenience of convicting an innocent to that of acquitting a culprit'. Later, the use of optimality principles to evaluate statistical procedures was evident in some of the key developments in mathematical statistics, such as estimation efficiency (Fisher, 1925) and the power of statistical tests (Neyman and Pearson, 1933), an approach that explicitly recognized the link between decision making and hypothesis testing. The broader connection between rational decision making and statistical inference was eventually formalized in great generality by Wald (1949), who is considered the founder of statistical decision theory. Current Bayesian decision theory stems from the theory of Savage (1954) and the seminal work of Raiffa and Schleifer (1961) and DeGroot (1970). Lindley (1985) presents the key concepts without complex mathematical logic, while Berger (1985), Robert (1994), and Bernardo and Smith (1994) provide more recent overviews and comparisons with alternative approaches.

2.6.2   Loss functions

In the setting of Section 2.2.1, each action identified, for every parameter value θ, a probability distribution on the set of possible outcomes. Generally, statistical decision problems are formulated in terms of choices among simpler actions which identify, for every parameter value θ, a single outcome. In this narrower definition, we can indicate the outcome that results from action a if θ is true by $a(\theta)$, and the resulting utility by $u(a(\theta))$.

Following Wald (1949), the statistical literature often formulates decision problems in terms of the opportunity loss (or 'regret') associated with each pair (θ, a) by defining a loss function

$$L(\theta, a) = \sup_{a(\theta)} u(a(\theta)) - u(a(\theta)), \qquad (2.20)$$

where the supermum is taken over the outcomes of all possible actions for fixed θ. The loss function $L(\theta, a)$ is the difference between the utility of the outcome of action a for state θ, and the utility of the outcome of the best action for that state (or the lowest upper bound if the best action cannot be achieved). The expected loss is

$$\mathcal{L}(a) = S_p - \mathcal{U}(a), \qquad (2.21)$$

where

$$S_p = \int_\Theta \sup_{a(\theta)} u(a(\theta)) p(\theta) d\theta$$

is independent of a. Therefore, the action minimizing the expected loss is the same as the action maximizing expected utility. In statistical analyses, loss functions are often stated directly, without reference to underlying consequences or utility functions.

2.6.3 Implications of the expected utility principle for statistical reasoning

Bayesian approaches are based on the notion that the distribution used in computing $\mathcal{U}$, or equivalently $\mathcal{L}$, at any stage in the decision process should make use of all knowledge available at the time of the decision, and that knowledge is incorporated by conditioning, that is, using Bayes' rule. Specifically, after observing data y, the prior distribution $p(\theta)$ is updated to obtain a posterior distribution,

$$p(\theta|y) = \frac{p(\theta)p(y|\theta)}{\int_\Theta p(\theta)p(y|\theta)d\theta}. \qquad (2.22)$$

An optimal choice a^* is one that maximizes expected utility – formally,

$$a^*(y) = \arg \max \int_\Theta u(a(\theta)) p(\theta|y) d\theta. \qquad (2.23)$$

The action that minimizes this expression will depend on y. So this procedure implicitly defines a decision rule for translating experimental evidence into actions.

This approach can be used to address statistical inference problems irrespective of data type, context, likelihood function, and current knowledge. Solutions are based on basic principles that are simple to state, understand, and refute if deemed inappropriate. Obstacles to implementation can occur, and simplifying assumptions are almost always needed in practice. However, in statistical applications, difficulties seldom arise because the conceptual framework is not sufficiently powerful. Specifying prior, likelihood, loss, and cost of experimentation is sufficient to characterize the statistical approach to be used. These elements can be stated explicitly and are common to all applications.

An immediate consequence of (2.23) is that observed experimental evidence affects Bayesian decision making only to the extent to which it is captured in the posterior $\pi(\theta|y)$, or equivalently by the likelihood $p(y|\theta)$. If sufficient statistics are available, they alone need to be considered in decision making. This principle is called called the likelihood principle (Berger and Wolpert, 1988). A controversial example of experimental evidence that does not enter the likelihood, but does affect frequentist inference, is provided by stopping rules that depend on the data but are independent of the state of nature (Berger, 1985; Bernardo and Smith, 1994). This controversy is critical in randomized clinical trials, which it may be ethical to stop when one of the treatments is manifestly superior (Berger and Berry, 1988).

Nuisance parameters are parameters that are required to specify a realistic model, but are not the primary object of investigation. It is difficult to develop general frequentist approaches in the presence of nuisance parameters. For example, the so-called most powerful tests are not generally applicable with nuisance parameters. This has generated a variety of *ad hoc* methods and approaches, and an ongoing controversy. From a decision-theoretic standpoint, a nuisance parameter can be defined as a parameter that appears in the sampling distribution but not in the loss function. For example, let $\theta = (\gamma, \nu)$, where ν is the nuisance parameter and γ the parameter entering the loss function, that is, $L(\theta, a) = L(\gamma, a)$ for every θ. The expected loss can be rewritten as

$$\mathcal{L}(a) = \int_\Gamma L(\gamma, a(x)) \int_V p(\gamma, \nu|x) d\nu d\gamma = \int_\Gamma L(\gamma, a(x)) p(\gamma|x) d\gamma.$$

The nuisance parameter ν is handled by averaging, as with all other uncertainties in expected utility theory. This provides a general and unified approach to statistical analyses in presence of nuisance parameters.

Multistage decision problems are common in medical applications, in which screening decisions feed into diagnostics decisions which in turn feed into treatment decisions. An example is discussed in Chapter 5. Successive updating of knowledge using Bayes' rule permits the modeling of accruing evidence, and also the prediction of future states of knowledge. This, in conjunction with backward induction algorithms, has constituted the basis for developing complex decision making strategies that adapt to new and accruing evidence (DeGroot, 1970).

In the remainder of this section we consider, very briefly, three examples of decision-theoretic statistical inference: hypothesis testing, classification in medical diagnosis, and sample size determination.

2.6.4   Hypothesis testing

Hypothesis testing problems have traditionally been among the earliest to be approached as decision problems (Neyman and Pearson, 1933). The action space usually includes endorsing the null hypothesis, specified as a subset of Θ, endorsing an alternative hypothesis, and possibly no endorsement or the requirement that more experimentation be carried out. The meaning of endorsing a hypothesis varies with the context, but in this paragraph and the next it means acting on the provisional assumption that the hypothesis is correct. The hypothesis testing framework is useful when experimentation is carried out to directly answer a specific dichotomous question. An example, based on soil samples, is the question 'Does this soil contain more than the allowed 0.004% of toxic compound A, so that reporting to the local authorities is required?'. Medical inquiry is usually more complex, as investigation of the strength of relationships and effects is critical, and experimental data serve multiple uses.

Hypothesis testing offers the opportunity to see decision-theoretic inferential ideas at work in an elementary setting. In the simplest formulation, Θ consists of only two points: θ_0 (the null hypothesis), and θ_1 (the alternative hypothesis), with prior probability $p(\theta_0) = p_0 = 1 - p(\theta_1)$. The action space includes actions a_0: 'endorse the null hypothesis' and a_1: 'endorse the alternative hypothesis'. This formulation, in both Bayesian and frequentist decision theory, is in contrast with Fisher's significance testing, which is concerned with measuring the extent to which data support a null hypothesis, separate from the consequences of rejecting that hypothesis.

All loss functions in this case can be described by two-way tables as in Table 2.11. The zeros on the diagonal are because loss functions are constructed so that there is at least one action with zero loss for every parameter value. The off-diagonal elements reflect the relative disadvantages of the two possible incorrect decisions.

The expected losses of the two actions are

$$\mathcal{L}(a_0) = L_1 p(\theta_1|y),$$
$$\mathcal{L}(a_1) = L_0 p(\theta_0|y),$$

and the Bayes rule is to choose a_0 whenever $\mathcal{L}(a_0) < \mathcal{L}(a_1)$, that is,

$$L_1 p(\theta_1|y) < L_0 p(\theta_0|y)$$

and a_1 otherwise. This is a formal restatement of Condorcet's opinion, discussed in Section 2.6.1.

Table 2.11 Loss function for testing a simple hypothesis against a simple alternative.

	a_0	a_1
θ_0	0	L_0
θ_1	L_1	0

Rewriting the inequality above by replacing Bayes' rule for posterior probabilities gives

$$\frac{p(y|\theta_0)}{p(y|\theta_1)} > \frac{L_1}{L_0}\frac{p_1}{p_0}. \tag{2.24}$$

The left-hand side is the so-called Bayes factor in favor of the null hypothesis. It measures the strength with which the experimental evidence supports the null hypothesis as compared to the alternative, and it summarizes the data for the purpose of this decision. The Bayes rule is to accept the null hypothesis when the Bayes factor is sufficiently large. The cutoff is determined based on the prior probability of the null hypothesis and the relative disutilities of the two possible incorrect decisions.

In this simple formulation, there is a close analogy between the decision function induced by the Bayes rule and the most powerful tests (Neyman and Pearson, 1933), which are in fact of the form (2.24). In frequentist formulations, however, the cutoff is chosen by controlling the significance level, that is, the probability of correctly endorsing the hypothesis in repeated experiments. In practice, this is frequently set to conventional values such as 0.95 or 0.99. This is an arbitrary choice. One can view expression (2.24) as one way of making explicit the implication of this choice.

More pronounced differences between Bayesian and frequentist hypothesis testing emerge in less stylized formulations. For example, if θ is a real-valued parameter and the null hypothesis is $\theta \in \Theta_0$, say $\Theta_0 = (\theta_0 - \epsilon, \theta_0 + \epsilon)$, the Bayes rule becomes

$$L_1 (1 - p(\Theta_0|y)) < L_0 \, p(\Theta_0|y)$$

or

$$\frac{\int_{\Theta_0} p(\theta)p(y|\theta)d\theta}{\int_{\Theta\setminus\Theta_0} p(\theta)p(y|\theta)d\theta} > \frac{L_1}{L_0}.$$

The notation $\Theta \setminus \Theta_0$ indicates the set of points in Θ but not in Θ_0. The expression above still reflects the general flavor of Condorcet's view, but is not consistent with typical frequentist approaches, except in very special cases.

Hypothesis testing is often taken to be a general paradigm of scientific inquiry, sometimes with deleterious effects on the progress of medical science and the quality of medical care. For example, confining the reported results of a medical study to the results of a hypothesis test is tantamount to identifying study investigators with DMs. Formally, this is reflected by the fact that choosing the cutoff for rejection is equivalent to setting priors and losses in a decision making framework. This identity of investigators and DMs is hardly appropriate in medicine, where physicians and patients will need to put the study information in the context of their knowledge, constraints, and values to reach their own decisions. Better reporting should be aimed at facilitating downstream decision making, for example, by reporting parameter estimates with associated errors, likelihoods, or sufficient statistics for reasonable models. Another example of far too great a reliance on the hypothesis testing paradigm is publication bias. This arises as a result of the practice of scientific journals of selectively publishing studies with statistically significant results, and therefore providing the scientific community and the public with a misleading view of the evidence.

2.6.5 Medical diagnosis

An important decision problem in medicine is diagnosis. When a single condition is considered, we can stylize diagnosis as the task of assigning an individual to one of two categories, based on a series of tests, covariates, or other information. Then there are two actions: a_1 is 'diagnosis of disease' and a_0 is 'diagnosis of no disease'. The structure of the loss function is analogous to Table 2.11. Normally, diagnosis is not an end in itself, but rather a means of assisting later treatment decisions. The losses L_0 and L_1 will reflect the consequence of the incorrect decisions that may stem from an incorrect diagnosis. These concepts can be extended to the case of more than two (say, d) possible diagnoses. Then there will be d actions. The loss function will be a $d \times d$ matrix with elements L_{ab}, with $a = 1, \ldots, d$ and $b = 1, \ldots, d$. Sometimes modeling of downstream decisions will be complex and require a multistage model, as in Chapter 5.

A simple situation is one in which the DM has data available on patient characteristics x and correct diagnosis y for a sample of individuals, and wishes to diagnose an additional individual, randomly drawn from the same population. One approach to classification is to develop a statistical prediction model $p(y|x, \theta)$ for the true disease status y given patient characteristics x and parameters θ; this could be a binary regression model, a Bayes network, a neural network, or any one of a variety of statistical models. See Duda et al. (2001) for a comprehensive review.

If x^* are the observed features of the new individual to be diagnosed, and y^* is the individual's unknown disease state, then the optimal decision is to choose a diagnosis of no disease if

$$L_1 p(y^* = 1|y, x, x^*) < L_0 p(y^* = 0|y, x, x^*).$$

It is useful to reexpress this condition in terms of the prediction model. The patient is a random draw from the same population that was used for training the model, which implies that x^* and y^* are conditionally independent of x and y given θ. Using this and the law of total probabilities,

$$p(y^* = 1|y, x, x^*) = \int_\Theta \pi(y^* = 0|x^*, \theta) p(\theta|x, y) d\theta.$$

Substituting this into the optimality condition, we can rewrite the diagnosis rule as

$$\frac{\int_\Theta p(y^* = 0 | x^*, \theta) p(\theta | x, y) d\theta}{\int_\Theta p(y^* = 1 | x^*, \theta) p(\theta | x, y) d\theta} > \frac{L_1}{L_0}.$$

This decision-theoretic diagnosis rule incorporates inference on the population model, uncertainty about population parameters, and the relative disutilities of misclassification errors.

2.6.6 Choosing the number of subjects in a study

Our last example concerns the choice of the sample size for an experiment. Specifically, we will consider the situation in which the DM has first to select a fixed sample size n, then observe a sample $y_1, \ldots, y_n$ and finally make a decision based on the data. We illustrate a decision-theoretic approach that models formally the view that the value of an experiment depends on the use that is planned for the results, and in particular on the decision that the results must help make. Raiffa and Schleifer (1961) pioneered Bayesian treatment of the optimal sample size problem. Textbook references on Bayesian optimal sample sizes include DeGroot (1970) and Berger (1985). For a review of Bayesian approaches to sample size determination, see Pham-Gia and Turkkan (1992), Adcock (1997), and Lindley (1997); for applications to medical studies see, for instance, Yao *et al.* (1996) and Tan and Smith (1998).

To choose a sample size for a study one needs to weigh the trade-off between carrying out a larger study and improving the final decision with respect to some criterion. This can be achieved in two ways: either by finding the smallest sample size that guarantees, say, a specified expected utility, or by explicitly modeling both the utility and the cost of experimentation. Formalizing this problem, we can set as our overall criterion a loss function of the form

$$L(\theta, a) + C(n), \tag{2.25}$$

where the term $L(\theta, a)$ is the loss function for the statistical decision problem that will follow collection of the data, while the function $C(n)$ represents the cost of experimentation as a function of the sample size.

Solving for the optimal sample size requires two steps. First, we need to find the optimal statistical decision a^* after the data are collected, for all possible data sets. This optimal decision will have an expected loss $\mathcal{L}_n(a^*)$, typically decreasing in n. Then we substitute this into (2.25) to find the sample that minimizes

$$\mathcal{L}_n(a^*) + C(n). \tag{2.26}$$

This is an instance of solving multistage decision problems by backward induction, an approach described in more detail in Chapter 5. For example, if the statistical decision problem is a simple hypothesis testing problem like the one described in Section 2.6.4, the expected loss associated with the optimal decision is $L_1 p(\theta_1|y)$ if a_0 is optimal and $L_0 p(\theta_0|y)$ if a_1 is optimal. So we can write

$$\mathcal{L}_n(a^*) = \max\{L_1 p(\theta_1|y), L_0 p(\theta_0|y)\}.$$

To determine the sample size for a study we can graph this criterion as a function of n, use it as an effectiveness measure in a cost-effectiveness analysis of alternative study designs, or formally minimize expression (2.26).

3

Simulation

3.1 Summary

Simulation is a computation method for exploring the implications of probabilistic modeling and variability in a wide variety of settings. It consists of generating artificial samples that are roughly consistent with specified probability distributions. The study of the samples is often much simpler than the study of the distributions originating those samples. For example, marginalization over high dimension, and transformations of variables of arbitrary form become trivial. In realistic medical decision models, random components are of high dimension, can be interconnected in complex ways, and vary over time. The properties of the resulting models are often analytically intractable, and must be investigated using numerical methods.

An efficient strategy is to set up an encompassing simulation-based framework for all computations that are required by the development, validation, and use of a decision model. This allows one to address in an integrated way four tasks that are often thought of as separate: (i) making inferences about the parameters of complex models; (ii) generating artificial cohorts of patients corresponding to different candidate interventions, and computing summaries of interest, such as expected utilities or cost-effectiveness ratios; (iii) performing sensitivity analysis of results with respect to input parameters and assumptions; and (iv) facilitating the search for optimal strategies in high-dimensional spaces. These aspects are considered in turn in this chapter, using three case studies.

1. *Sensitivity and specificity estimation in the absence of a gold standard.* We consider a study in which a new test for tuberculosis is to be evaluated, but it is not feasible to ascertain the true disease status of the patients tested. Instead, the new test is evaluated against a standard test (van Meerten et al., 1971). Elaborating on the methodology developed by Hui and Walter (1980), we will show how to make inferences about sensitivity and specificity. The goals of this example are to introduce inference for multiparameter problems, and to illustrate Markov chain Monte Carlo exploration of complex posterior distributions.

2. *Chronic disease modeling revisited.* In Chapters 1 and 2 we discussed the μSPPM, a miniature model for prediction of the natural course of stroke-related events (Parmigiani *et al.*, 1997). Here we will revisit the μSPPM to illustrate both points (ii) and (iii) above.

3. *A two-phase design in screening for a rare disease.* An effective strategy for estimating the prevalence of a rare illnesses is to identify individuals that are at high risk using either an inexpensive screening test or risk factors, and then test a proportion a_1 of the patients who screened positively and a proportion a_2 of the patients who screened negatively using a more accurate test (Shrout and Newman, 1989; Erkanli *et al.*, 1998). We examine the efficiency of the design as a function of the two proportions a_1 and a_2 and use this example to illustrate an approach to random sampling of candidate designs.

The use of simulation methods as a research tool originated with work by the Manhattan Project during the Second World War. The term Monte Carlo, often used to designate simulation-based computational techniques, was introduced by Metropolis during that time (Metropolis, 1987). After over 50 years of application in most fields of science and technology, the literature is extensive. A general discussion of random number generation is provided by Niederreiter (1992) and extensive technical details are given by Knuth (1997). More statistically oriented treatments include Ripley (1987) and Gentle (1998). Markov chain Monte Carlo (MCMC) methods for Bayesian inference are discussed by Gilks *et al.* (1996) and Gamerman (1997). More technical aspects of MCMC are developed by Robert and Casella (1999).

3.2 Inference via simulation

3.2.1 Inference on sensitivity and specificity without a gold standard

In this section we reconsider the estimation of specificity and sensitivity, examining a more complex situation than that analyzed in Section 1.4. Here, a new diagnostic test, in our example the Tine test for tuberculosis, is to be evaluated, but it is not feasible to ascertain the true disease status of the patients tested. Instead, the new diagnostic test is evaluated against a standard test, in our case the Mantoux test, with its own errors. Ignoring the error rates of the standard test can lead to overestimation of the error rates of the new test. Van Meerten *et al.* (1971) give an example of this, pointing out that the errors of computer diagnosis are inflated when it is evaluated against a physician's diagnosis which is taken as correct.

Table 3.1 displays the data that we will analyze. Patients in two populations are cross-classified according to the outcome of the Mantoux and Tine tests. The two populations are a southern US school district and the Missouri state sanatorium. Sampling patients in two populations that are likely to have a different prevalence of tuberculosis mimics the ideal situation, in which

Table 3.1   Cross classification of patients according to the outcome of the Mantoux and Tine tests for tuberculosis, in two populations: 1) a southern U.S. School district and 2) the Missouri State Sanatorium.

	School district		Sanatorium	
	Tine Test +	Tine Test−	Tine Test +	Tine Test−
Mantoux Test +	14	4	887	31
Mantoux Test −	9	528	37	367

we would sample patients known to have disease and patients known to be free of disease. A study categorizing a single population of patients based on the outcomes of the two tests cannot provide information on the error rates of both. However, if data are available on two populations with different disease prevalences, one can estimate the error rates of both tests, as well as the prevalences in both populations. This interesting result is discussed in a seminal paper by Hui and Walter (1980). Here we will discuss a Bayesian version of their model.

We begin with some notation: β_1 and β_2 are the sensitivities of the two tests. The subscript 1 indicates the Mantoux test and the subscript 2 indicates the Tine test. Similarly, α_1 and α_2 are the two specificities. Let us now consider the two samples of patients, each cross classified into a 2×2 table according to the outcome of the two tests. This gives us a total of 8 cells (see in Table 3.1). For the entries in the table for population i, let us use the notation n_i^{++}, n_i^{+-}, n_i^{-+} and n_i^{--}, where n_i^{+-} is the number of patients that tested positive on test 1 and negative on test 2, and so forth. Here i can be either 1 or 2. The prevalence in each population is some unknown value π_i.

We will also assume that the outcomes of the two tests are independent given disease status. This means assuming that if a patient is healthy, a correct diagnosis with test 1 does not change the probability of a correct diagnosis with test 2, and likewise if the patient is diseased. In other words, there are no patients that are comparatively difficult or easy to diagnose for both tests, and the reasons why tests make mistakes are unrelated. In symbols,

$$P(T_1, T_2|D) = P(T_1|D)P(T_2|D),$$

where T_i are test results, and D is disease status. This may be a unrealistic assumption in some situations, but for now let us pretend it is correct.

To write the likelihood function, we need to understand how the probability of belonging to each of the cells depends on the parameters of interest. This is similar to what was done in Table 1.3, when working out the marginal probability of testing positive as a function of prevalence, sensitivity and specificity. Now this needs to be done for both populations and both tests. For example, the probability that an individual in population 2 will test positive on test 1 and negative on test 2 is

$$
\begin{aligned}
P(T_1+, T_2-) &= P(D+)P(T_1+, T_2-|D+) + P(D-)P(T_1+, T_2-|D-) \\
&= P(D+)P(T_1+|D+)P(T_2-|D+) \\
&\quad + P(D-)P(T_1+|D-)P(T_2-|D-) \\
&= \pi_2 \beta_1 (1 - \beta_2) + (1 - \pi_2)(1 - \alpha_1)\alpha_2,
\end{aligned}
\tag{3.1}
$$

where the first equality follows from the law of total probabilities, the second from the conditional independence assumption, and the last is from the notation.

Using a similar argument for the other seven possibilities, and considering that the patients are a random sample from the corresponding populations, we obtain the likelihood function

$$
\begin{aligned}
p(n_1^{++}, &n_1^{+-}, n_1^{-+}, n_1^{--}, n_2^{++}, n_2^{+-}, n_2^{-+}, n_2^{--} | \beta_1, \beta_2, \alpha_1, \alpha_2, \pi_1, \pi_2) \\
&= [\pi_1 \beta_1 \beta_2 + (1 - \pi_1)(1 - \alpha_1)(1 - \alpha_2)]^{n_1^{++}} \\
&\quad \times [\pi_1 \beta_1 (1 - \beta_2) + (1 - \pi_1)(1 - \alpha_1)\alpha_2]^{n_1^{+-}} \\
&\quad \times [\pi_1 (1 - \beta_1)\beta_2 + (1 - \pi_1)\alpha_1(1 - \alpha_2)]^{n_1^{-+}} \\
&\quad \times [\pi_1 (1 - \beta_1)(1 - \beta_2) + (1 - \pi_1)\alpha_1\alpha_2]^{n_1^{--}} \\
&\quad \times [\pi_2 \beta_1 \beta_2 + (1 - \pi_2)(1 - \alpha_1)(1 - \alpha_2)]^{n_2^{++}} \\
&\quad \times [\pi_2 \beta_1 (1 - \beta_2) + (1 - \pi_2)(1 - \alpha_1)\alpha_2]^{n_2^{+-}} \\
&\quad \times [\pi_2 (1 - \beta_1)\beta_2 + (1 - \pi_2)\alpha_1(1 - \alpha_2)]^{n_2^{-+}} \\
&\quad \times [\pi_2 (1 - \beta_1)(1 - \beta_2) + (1 - \pi_2)\alpha_1\alpha_2]^{n_2^{--}}.
\end{aligned}
\tag{3.2}
$$

To see why data on a single population are not sufficient to make inferences on β_1, β_2, α_1, α_2, and π_1, imagine having a massive amount of data from population 1. Then the values of the expressions in square brackets in the first four. lines of the right-hand side would be known very accurately, while those in the last four lines would not be known at all. Could we translate this knowledge into knowledge about β_1, β_2, α_1, α_2, and π_1? Both the first four and last four expressions in square brackets are entries in a 2×2 table and so they add up to 1. We have three independent quantities to solve for five unknowns. So there will be many values of β_1, β_2, α_1, α_2, and π_1 that are consistent with the cell probabilities. In statistical terms, the parameters are not *identifiable* based on this type of experiment. When we consider two populations, however, if the sensitivities and specificities do not vary across populations, we add one parameter (π_2) and three independent cell probabilities. So we now have six independent expressions in square brackets to solve for six unknowns.

There is one additional difficulty. In general, there will be two ways to solve our system of six equations in six unknowns. If α_i, β_i, and π_i is a solution, then so is $1 - \alpha_i$, $1 - \beta_i$, and $1 - \pi_i$. That is natural, because we do not know the disease status of the patients, so a world with exchanged D+ and D−, and 'reversed' sensitivities and specificities, would give rise to the same empirical evidence. Suppose we observe many patients who are positive to both tests. In the first case, the conclusion is that there are many diseased patients, and the tests are detecting them. In the second, reverse case, it is that there are many healthy patients, and the tests are misclassifying many of them as diseased. We cannot distinguish empirically between the two. In most cases, however, it is reasonable to rule out the 'reverse' case based on prior knowledge about one of the two tests. A constraint that is sufficient to rule out the 'reverse' case is that $\alpha_1 + \beta_1 > 1$. If test 1 is a standard test, this is typically a quite reasonable assumption.

In a Bayesian analysis, it is convenient to specify constraints by setting the prior probability or density to zero outside the region of the constraint. In our example we will have the additional constraint $\pi_1 < \pi_2$. For this analysis we specify a uniform prior for all parameters in the region satisfying the constraints.

3.2.2   Sampling from the posterior distribution

The joint posterior distribution of the parameters β_1, β_2, α_1, α_2, π_1, and π_2 is the same as the likelihood function in expression (3.2), because we are using uniform prior distributions. Studying the implication of expression (3.2) is not as simple as in Section 1.2.2. First the posterior distribution is of higher dimension, although we are still interested in studying one or two dimensions at a time. Second, the functional form of the posterior distribution is not one that is readily available for direct simulation from standard statistical packages. Third, there are two constraints on the parameters.

Here we describe a simulation strategy for drawing inferences based on the joint posterior distribution of the parameters. We take two important facts for granted. First, we can evaluate interesting summaries of a complex posterior distribution, such as marginal means, probability intervals, and probabilities of arbitrary subregions, by drawing samples from this distribution, and then looking at the corresponding sample means, quantiles, and frequencies. Second, with a little bit of care, we can generate a sample from a high-dimensional joint distribution by iteratively generating samples from the conditional distributions of each dimension given all others. We will return to both these issues in Section 3.2.3.

A key simplifying factor here is that we can think of our inference problem as an incomplete version of a larger but simpler analysis problem, in which we know the true disease status of every subject in the sample. In this larger problem, it is straightforward to make inferences about prevalences, sensitivities and specificities – in fact we are in a very similar situation to the binomial inference problem of Section 1.2.2. In particular, we can easily generate values from the distributions of all the parameters conditional on the disease status of the subjects. Our strategy will therefore be to investigate the larger problem, including the true disease status of the subjects, treating the disease status variables of every patient in the sample as unknowns and simulating them. Each cycle of this simulation approach will be then in two simple phases:

1. Simulate disease status conditional on parameters.
2. Simulate parameters conditional on disease status.

Cycles are iterated until a representative sample is gathered. The sequence of sampled values is a random process. Values at each cycle are dependent on values of the previous cycle. The first cycle requires a choice of plausible initial values of the parameters. For example, we can take the empirical estimates assuming everyone in the sanatorium is diseased and everyone in the school district is healthy.

To implement our data augmentation scheme, consider phase 1 first, and assume we are past the initial iteration. The disease status of all subjects is set to the value generated at the previous iteration. Denote by m_t^+ the numbers of diseased individuals who tested positive in test t, and by m_t^- the number of diseased individuals who tested negative. Let, q_t^+ and q_t^- represent the analogous counts for healthy individuals. Conditional on disease status we can consider each test separately, as they are conditionally independent, and sensitivity and specificity separately. Each subproblem is a simple binomial problem like that of Section 1.4.4, with uniform prior, so we obtain

$$\alpha_t \sim \text{Beta}\left(\frac{q_t^- + 1}{q_t^+ + q_t^- + 2}, \frac{q_t^+ + 1}{q_t^+ + q_t^- + 2}\right),$$

$$\beta_t \sim \text{Beta}\left(\frac{m_t^+ + 1}{m_t^+ + m_t^- + 2}, \frac{m_t^- + 1}{m_t^+ + m_t^- + 2}\right)$$

for $t = 1, 2$, subject to the constraint $\alpha_1 + \beta_1 > 1$. Similarly, the distributions of prevalence in the two populations are easily updated. If r_i and s_i are the numbers of diseased and healthy individuals in population i,

$$\pi_i \sim \text{Beta}\left(\frac{r_i + 1}{r_i + s_i + 2}, \frac{s_i + 1}{r_i + s_i + 2}\right)$$

subject to the constraint $\pi_1 < \pi_2$. A simple approach to accounting for the constraint is to sample parameter values from their unconstrained distribution, check that they satisfy the constraint, and if they do not then replace them with the parameter values from the previous iteration. Another approach is to keep generating values until the constraint is satisfied. This is generally less efficient, although more intuitive.

For phase 2, assume now that all the parameters above are fixed to the values just generated, and update the disease status of each individual in the sample. There are eight types of individuals, corresponding to each of the cells in Table 3.1. They can all be analyzed similarly, so we only look at one cell. For example, for an individual in the sanatorium group, testing positive in test 1 and negative in test 2, the probability of disease is, using Bayes' rule,

$$\frac{\pi_2 \beta_1 (1 - \beta_2)}{\pi_2 \beta_1 (1 - \beta_2) + (1 - \pi_2)(1 - \alpha_1)\alpha_2}.$$

Because all individuals in the cell are the same in this regard, we can generate the summary ms and qs directly, from the appropriate binomial distributions, rather than simulating subjects individually.

This iterative updating procedure is a special case of a general class of algorithms known as Markov chain Monte Carlo algorithms, described in additional detail in Section 3.2.3. The idea of solving a bigger but simple problem by simulating the missing parts exemplifies a powerful computational strategy called data augmentation (Tanner and Wong, 1987).

Sampled parameter values from the Markov chain are shown in Figure 3.1. The quantiles of the marginal posterior distributions of the sensitivities and specificities are as follows:

	2.5% quantile	97.5% quantile
α_1	0.981	0.998
α_2	0.968	0.991
β_1	0.951	0.977
β_2	0.955	0.979

Because of the relatively large sample sizes of the study, the use of a flat prior distribution, and the low values of the likelihood function outside the region of the constraints, the results are quite close to those of a likelihood analysis

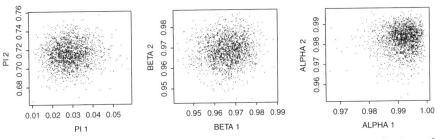

Figure 3.1 Sample of values from the posterior probability distribution of prevalences (left), sensitivities (center), and specificities (right) in the tuberculosis example.

based on a normal approximation to the likelihood function, reported by Hui and Walter (1980).

To illustrate the effect of the constraints and the use of an informative prior distribution, we now consider a hypothetical example, based on the same sampling design. The data are shown in Table 3.2. Sample sizes are smaller than those of the tuberculosis example. Also, three of the cells are empty, which creates a complication in maximum likelihood analysis, as the estimates are likely to be on the boundary of the parameter space.

A sample from the posterior distribution of the parameters is shown in Figure 3.2. The left-hand panel emphasizes the effect of the constraint $\pi_1 < \pi_2$, making it impossible for samples from the posterior distribution to fall below the diagonal line. The center and left-hand panels illustrate how the mode of the posterior distribution, which corresponds to the region of highest density of dots in the figures, is on the boundary of the square for α_1, α_2, and β_1. So the mode of the posterior is 1, as is the maximum likelihood estimate. However, in view of the substantial remaining uncertainty about all three of the parameters, a more complete exploration of the posterior distribution (or the likelihood function) is appropriate. Sampling provides a convenient framework for such exploration as one does not have to rely on

Table 3.2   Cross-classification of patients in two populations according to the outcome of the standard and new diagnostic tests.

	Population 1		Population 2	
	New Test +	New Test −	New Test +	New Test −
Standard Test +	1	0	8	4
Standard Test −	0	38	0	87

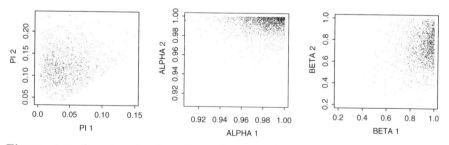

Figure 3.2   Sample of values from the posterior probability distribution of prevalences (left), sensitivities (center), and specificities (right) in the hypothetical example, without prior information on the standard test.

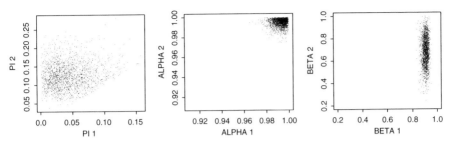

Figure 3.3 Sample of values from the posterior probability distribution of prevalences (left), sensitivities (center), and specificities (right) in the hypothetical example, with prior information on the standard test.

asymptotic approximation of likelihood and posterior distributions, and can handle automatically a very broad range of shapes.

In comparing a standard test to a new test, we may have the option of eliciting reliable expert knowledge on the sensitivity and specificity of the standard test. Here we assume that, based on the expert's knowledge, we can specify that α_1 is Beta(248, 2), and β_1 is Beta(225, 25). A sample from the new posterior distribution of the parameters is shown in Figure 3.3. The spread of α_1 and β_1 is much reduced and the posterior sample is now markedly affected by the prior. Because we do not know the true disease status of the patients, the sensitivities and specificities of the two tests are not independent. In particular, the spread of α_2 is noticeably affected by the more informative prior on α_1.

3.2.3   Markov chain Monte Carlo methods

The computational approach of Section 3.2.2 exemplifies a general approach to the investigation of posterior distributions using simulation. We argued that there are two key components to making it work successfully. First, we can replace the analytic calculation of posterior quantities of interest with summaries of simulated values. Second, when we cannot generate simulated values directly, we may still be able to break down a large problem into smaller and simpler pieces, and simulate one piece at the time.

The first component is based on the principle of Monte Carlo integration. Suppose we are interested in a parameter θ and we have a posterior distribution $p(\theta|\text{Data})$. We will suppress explicit reference to the dependence on the data in the rest of this section. Suppose we can generate a set of independent and identically distributed values $\theta_1, \ldots, \theta_M$ from p. Most quantities of the form

$$\int_{\Theta} h(\theta)p(\theta)d\theta$$

can be evaluated using the corresponding Monte Carlo average

$$\frac{1}{M} \sum_{m=1}^{M} h(\theta_m). \tag{3.3}$$

Here h is a general (although not completely arbitrary) function that captures the goal of the integration. For example, to evaluate a posterior mean take $h(\theta) = \theta$, that is, average the sampled points, and to evaluate the probability that θ is greater than (say) 3.1, take $h(\theta) = 1$ if $\theta > 3.1$ and $h(\theta) = 1$ otherwise, that is, take the sample frequency of values greater than 3.1. Densities can also be estimated, for example by simple kernel density estimation techniques available in most statistical packages. Because of the law of large numbers, the Monte Carlo average can be made arbitrarily close to the solution of interest by increasing M (Gentle, 1998).

In multiparameter problems, it is easy to evaluate distributions of arbitrary functions of the parameters. For example, in the context of Section 1.2.2, if we have a sample of values π_m, α_m, β_m, $m = 1, \ldots, M$, and we are interested in the positive predictive value π^+, we can obtain a random sample from the distribution of π^+ by simply evaluating equation (1.1) for each m, that is, by computing

$$\pi_m^+ = \frac{\pi_m \beta_m}{\pi_m \beta_m + (1 - \pi_m)(1 - \alpha_m)}, \qquad m = 1, \ldots, M. \tag{3.4}$$

We can then use Monte Carlo averages to summarize this distribution as well.

In practice, it is not always possible to generate independent samples of θs. The principle of Monte Carlo approximation, however, applies to a much broader class of sampling schemes. The critical aspect is that, for large values of M, each θ is sampled in the correct proportion. One way of achieving this is a random process such as that of Section 3.2.2. There, samples are not independent, in fact they constitute a Markov chain. For each value θ_m, including parameters and disease status summaries, our algorithm described a probabilistic mechanism for generating a new value θ_{m+1}, from a distribution implicitly defined by the list of steps in phases 1 and 2 on page 96. This random process converges to the correct stationary distribution, that is, it eventually stabilizes and generates values of θs in the right proportions.

The sampling strategy of Section 3.2.2 is an example of Markov chain Monte Carlo, a widely applicable approach that, in the last decade, has revolutionized the way statistical modeling is understood and implemented. Excellent introductions to MCMC include Gilks et al. (1996) and Gamerman (1997). More technical aspects are developed by Tierney (1994) and Robert and Casella (1999). Without attempting to cover the vast array of alternative algorithms, we briefly describe some of the most popular options for constructing Markov chains to explore posterior distributions. Some technical conditions on p are necessary for MCMC to work, among which that it should be possible to move from any value θ to any other value θ'. Details can be found in the references above.

One of the oldest ways of constructing a Markov chain with a specified stationary distribution is the Metropolis–Hastings algorithm (Metropolis et al., 1953; Hastings, 1970). We are interested in generating a sample of values θ_m, $m = 1, \ldots, M$, from $p(\theta)$. We cannot do so directly. Instead, we choose a starting point and move from there in a way that is driven by the probability distribution p. If we are at θ_m after m iterations, we generate the next value in two steps: we first propose a value θ^* and then either accept it as the new θ_{m+1} or reject it and retain the old one. We propose a new value randomly by generating it from a distribution $p^*(\theta^*|\theta_m)$, called the proposal distribution. Then we accept the candidate point randomly with probability

$$\min\left(1, \frac{p(\theta^*)p^*(\theta_m|\theta^*)}{p(\theta_m)p^*(\theta^*|\theta_m)}\right).$$

If the candidate is accepted, $\theta_{m+1} = \theta^*$. If not, $\theta_{m+1} = \theta_m$.

The proposal distribution is usually chosen so that it is easy to generate from it. For example, it may be a multivariate normal centered at the current value θ_m and having a fixed covariance matrix. If the proposal distribution is symmetric, that is, if $p^*(\theta'|\theta) = p^*(\theta|\theta')$ for every couple (θ, θ'), then the acceptance probability above simplifies to

$$\min\left(1, \frac{p(\theta^*)}{p(\theta_m)}\right),$$

the beautifully simple idea that was at the core of the original Metropolis algorithm (Metropolis et al., 1953). Generally, a very wide set of proposal distributions will lead to a Markov chain with the right stationary distribution p. Choosing the proposal distribution carefully can, however, substantially affect the efficiency of a chain and, in complex problems, can determine whether a sampling scheme may be ultimately implemented or not.

In posterior inference, it is often the case that we can easily evaluate pointwise both the prior and the likelihood, and therefore the numerator of the posterior distribution as expressed by Bayes' rule. However, evaluating the denominator can involve far more work, especially in high-dimensional cases. Because the denominator is independent of θ, it cancels out in the computation of the acceptance probabilities, and it is not required in the Metropolis–Hastings scheme. This is a critical aspect in complex applications.

In multidimensional problems, instead of updating all components of θ simultaneously, it is often more efficient to consider one subset at a time. For illustration, consider the case in which $\theta = (\alpha, \beta)$, where α and β are two subsets of parameters, for example the sensitivities and specificities in Section 3.2.2. Then each iteration consists of two phases:

1. Determine α_{m+1} given α_m and β_m.
2. Determine β_{m+1} given α_{m+1} and β_m.

Each phase proceeds according to a Metropolis–Hastings update as described above. This requires two separate proposal distributions for α and β.

This logic enables one to break down complex problems into simpler components and carry out the simulation for one component at a time given the others. It is ideal for models that have multiple interconnected sections, or multiple levels, as we will see in Chapter 4. A classic implementation of this general idea is the Gibbs sampler (Geman and Geman, 1984; Gelfand and Smith, 1990). In a Gibbs sampler, phases 1 and 2 above are as follows:

1. Draw α_{m+1} from $p(\alpha|\beta_m)$.
2. Draw β_{m+1} from $p(\beta|\alpha_{m+1})$.

These distributions are derived directly from the joint posterior distribution, and are called full conditional posterior distributions. The Gibbs sampler could be viewed as a special case of the Metropolis–Hastings algorithm in which the distributions above are used as proposal distributions. The proposed values are always accepted because, as can easily be checked, the acceptance probability is identically 1.

Starting values of the MCMC algorithm do not theoretically change the limiting distribution of a valid sampler, but still need to be chosen with some care, as they can affect efficiency substantially. Typically, draws are used for summarization only after discarding an initial section (termed the burn-in section) of the chain. In complex problems, it can be wise to run in parallel several chains from multiple starting values, to ensure that the answers are indeed unaffected by the choice of initial values. Choosing the length of the chain, the length of the burn-in, assessing whether the chain is likely to provide samples from the stationary distribution, and measuring Monte Carlo uncertainty around quantities of interest can be complex, and are discussed in Gilks et al. (1996).

A software package called BUGS (for Bayesian inference Using Gibbs Sampling) offers easy implementation of MCMC in a wide variety of statistical models (Gilks et al., 1994; Spiegelhalter et al., 1996). The so-called classic version of BUGS uses text-based model descriptions and a command-line interface. Versions are available for major computer platforms. A Windows version, WinBUGS, also has the option of a graphical user interface and has on-line monitoring and convergence diagnostics. BUGS is freely available from the site http://www.mrc-bsu.cam.ac.uk/bugs/welcome.shtml. BUGS manuals also serve as an excellent introduction to multilevel Bayesian modeling. CODA is a suite of S-plus/R functions for performing convergence diagnostics.

In addition, software for MCMC is available for a wide variety of specific applications. For example, Flexible Bayesian Modeling, or FBM (Neal, 1996), supports Bayesian analysis of regression and classification models based on neural networks and Gaussian processes, as well as Bayesian mixture models. It also supports a variety of Markov chain sampling methods, which may be specified by simple formulae representing priors and likelihoods.

3.3   Prediction and expected utility via simulation

3.3.1   Generating synthetic cohorts

The starting point of a simulation-based decision analysis is to generate synthetic individual-level data for various hypothetical cohorts, corresponding to each combination of actions and covariates. This process is sometimes called microsimulation. An early example in health care is the program MISCAN, developed to evaluate screening for occult disease on a population basis (Habbema et al., 1985).

Simulation proceeds on the basis of the predictive distributions $p_a(z|x, \text{Data})$ of outcomes z for subjects with covariates x when action a is taken. These can then be used for summarization to inform decision making. The outcome model $p_a(z|x, \theta)$ and the posterior distribution $p(\theta|\text{Data})$ can be used as intermediate steps to determine the predictive distribution via the equation

$$p_a(z|x, \text{Data}) = \int_\Theta p_a(z|x, \theta)p(\theta|\text{Data}) \, d\theta. \tag{3.5}$$

An efficient strategy for generating simulated samples of outcomes measures, and incorporating the uncertainty about the model parameters, is to set up a single simulation scheme to perform estimation, combination of information, and prediction. The basic idea is to sample from the joint distribution of z and θ, given the data, and then use only the simulated values for the outcome, which constitute a draw from the marginal predictive distribution of equation (3.5). In practice, this can be implemented by first generating a sample of θ, perhaps using an MCMC approach, and then, for each of the sampled values of θ, generating a sample of outcomes from the outcome model.

Using the resulting sample of outcomes for decision making is then a simple matter of summarization, which can be done using standard graphical tools such as boxplots or histograms. Important summaries such as expected utility, as in expression (2.10), or cost-effectiveness ratios can be evaluated by Monte Carlo averages. Predictions incorporate variability of model parameters.

The synthetic cohorts generated from equation (3.5) constitute a powerful and user-friendly way of representing and storing information about the disease under consideration and the relevant treatment option. These cohorts are the result of information from different sources and can incorporate substantial clinical expertise. When each source is used for the purpose to which it is best suited, these cohorts can be a repository of state-of-the-art information and be substantially more informative than any of the individual studies that contributed to their development.

3.3.2   Generating a synthetic cohort in the μSPPM

In our discussion in Sections 1.5.1 and 2.4.2 we presented a Markov process, represented by Figure 1.11, for describing the history of stroke-related events, and for studying the effects of stroke prevention interventions. Many of the summaries that are relevant to decision making can be written as closed-form expressions in terms of the unknown parameters, as exemplified in Section 2.5.2, and do not require simulation of cohorts. However, it is interesting to review the steps that are necessary to implement the synthetic cohort approach in this case.

Table 2.9 summarizes the sufficient statistics entering the posterior distributions of the λs, using the gamma distributions (1.28). We also consider an intervention that reduces the hazard of making a transition from the TIA state to the STROKE state by κ^1 in stenosis patients and by κ^0 in other patients. We assume that, based on uncertainty reported in the literature, κ^0 and κ^1 are normal with means 0.95 and 0.77, and standard deviations 0.05 and 0.03. Also, perioperative mortality is observed in 3% of patients in the published study, so the parameter ρ, the probability of surviving the procedure, is modeled as a beta random variable with parameters 97 and 3.

Using this information, we can implement a simple algorithm for generating the four synthetic cohorts corresponding to each combination of treatment and stenosis status. We need to iterate the following two steps:

1. Simulate the λs, κs and ρ.
2. Simulate outcome histories z given the λs, κs and ρ.

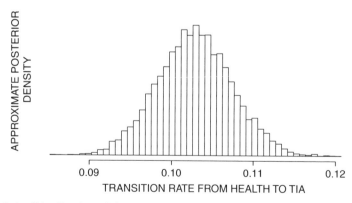

Figure 3.4   Distribution of the transition rate from HEALTH to TIA for covariate $x = 0$.

In our example, the distributions of the λs, κs and ρ are simple, and can be sampled directly, without resorting to MCMC. Figure 3.4 shows a sample from the gamma distribution of the transition rates from health to the TIA state. Similar samples can be obtained for the other transition rates. Simulation of patient histories is also straightforward and can be implemented along the following lines. Fix a cohort size M, a covariate level x, and an action (here we consider the treatment arm). For each individual in the cohort:

1. Generate the first transition, in two steps:
   - Generate the time in H, t_H, from $\mathrm{Exp}(\lambda_H^x)$.
   - Generate the direction of transition, choosing T with probability $\lambda_{HT}^x/\lambda_H^x$, S with probability $\lambda_{HS}^x/\lambda_H^x$, and D with probability $\lambda_{HD}^x/\lambda_H^x$.

2. Generate the second transition, if the first is not to D. This takes place again in two steps, depending on the destination state.

   If it is T, choose perioperative death with probability $1-\rho$, or survival with probability ρ. If survival, generate the time in T from $\mathrm{Exp}(\kappa^x\lambda_{TS}^x + \lambda_{TD}^x)$. Generate the direction of transition, choosing S with probability $\kappa^x\lambda_{TS}^x/(\kappa^x\lambda_{TS}^x + \lambda_{TD}^x)$, and D with probability $\lambda_{TD}^x/(\kappa^x\lambda_{TS}^x + \lambda_{TD}^x)$.

   If it is S, generate the time in S, t_S, from $\mathrm{Exp}(\lambda_S^x)$.

   If it is D, stop.

3. Generate the third transition, if the second transition is not to D. This step must be from S and proceeds as above.

These steps are repeated M times to generate all subjects in the cohort. A number of summarizations are now possible. For example the mean life expectancy in the cohort for individuals in the TIA cohort is 1.182, while the variance is 0.98. These results can be contrasted to those obtained without considering parameter uncertainty in the simulation of the cohort, which lead to a similar mean of 1.177, but a substantially reduced variance of 0.85. Additional summarizations will be illustrated in Section 3.4.1.

3.4   Sensitivity analysis via simulation

3.4.1   Probabilistic sensitivity analysis

Bayesian predictions produce answers that incorporate both the variability in the unknown patient histories z and the unknown model parameters θ. Often it is interesting to separate the two sources of variability in presenting the results of a decision analysis. For example, there is a growing recognition that reporting estimates of marginal cost-effectiveness ratios (MCERs) is necessary but not sufficient, and that describing their uncertainty with respect to model parameters, while still averaging with respect to individual variation, is useful in decision making (Gardiner et al., 1995; Bult et al., 1995; Manning et al., 1996; Stinnett and Mullahy, 1998). Imperfect knowledge of parameter values and assumptions can be handled in three somewhat complementary ways: marginalization, scenario sensitivity analysis, and probabilistic sensitivity analysis. Marginalization is the approach derived from expected utility and is discussed in Section 2.4.4. Scenario sensitivity analysis is based on drawing a list of interesting values of the input parameters, evaluating expected outcomes under all of them, and graphically or otherwise assessing the strength of the effect of the inputs on the outputs. It is especially useful when the scenarios correspond to alternative theories in a scientific controversy. It is also used often to examine sensitivity to the choice of prior distributions in Bayesian analyses. Scenario sensitivity analysis is subject to the fallacy of backsolving for input values that give desired output values. Therefore it is critical to disclose the process that led to the selection of specific scenarios and, when possible, to tie the scenarios to specific theories, studies, or models. An illustration is given in Section 6.3.4.

Probabilistic sensitivity analysis is based on drawing a sample from a distribution of θ, ideally a posterior distribution, or a distribution that reflects the state of knowledge about the parameters, on evaluating the expected outcomes for each draw, and on studying the resulting collection of sampled outcomes. Probabilistic sensitivity analysis by simulation has been advocated in the medical context by various authors, including Doubilet et al. (1985) and Critchfield and Willard (1986). Formally, it can be described in three steps:

1. Draw $\theta_1, \ldots \theta_M$ from $p(\theta|\text{Data})$, perhaps using MCMC.

2. For each $m = 1, \ldots M$, compute the expected utilities of interest by averaging over subjects in the cohort, but not over parameter values, that is, evaluate

$$\mathcal{U}_m(a) = \int_{\mathcal{Z}} u(z) p_a(z|x, \theta_m) dz.$$

3. Summarize:

   (a) the overall variability in $\mathcal{U}_m(a)$ due to imperfect knowledge of the parameters;

   (b) the dependence of $\mathcal{U}_m(a)$ on individual components of θ_m; and

   (c) the dependence among outcomes.

This approach to probabilistic sensitivity analysis naturally incorporates dependencies among components of θ. Examples of summarization strategies are given in Section 5.4.6.

Alternative ways of determining parameter distributions for probabilistic sensitivity analysis include assessment of expert opinion and resampling methods such as the bootstrap (Parmigiani et al., 1997).

3.4.2   Probabilistic sensitivity analysis of the MCER in the μSPPM

We return once again to the μSPPM and consider uncertainty about the results of Section 2.5.2. We study the distribution of cost, effectiveness, and the MCER implied by our posterior distribution on the parameters λs, κs and ρ. Support for considering the entire distribution of the MCER comes from the nature of comparisons between controversial health interventions. In many cases, there is the potential for a large advantage and yet a substantial probability of a small advantage. This leads to a pronouncedly skewed distribution for the MCER, which is difficult to summarize using simply the mean MCER and a confidence interval. In addition, simulation-based distributions can be used to construct probability intervals and other probability statements relevant for policy making. Examples include the probability that the intervention is more effective than standard care, and the probability that the cost-effectiveness ratio is lower than a given value of interest.

In the case of the μSPPM, we can derive a probability distribution for the cost, effectiveness, and MCER by simulating from the posterior distribution of model λs, κs and ρ, as was done in Section 3.3.2, and then evaluating the MCER of Section 2.5.2 for each set of parameters generated. We generated 10 000 samples for both actions and both stenosis values. From these draws we can derive distributions of life expectancies, and the distribution of the gain in life expectancy from carrying out the intervention rather than standard care. We can also evaluate the distributions of costs both within groups and comparing intervention and control. In the μSPPM, the treatment is costly in two ways: the initial amount and the prolonged sojourn in the TIA state, which is random.

Figure 3.5 shows the joint distribution of the increase in costs and increase in effectiveness associated with the intervention. This distribution contains all the information needed for cost-effectiveness analysis and can be thought of as the goal of an analysis of uncertainty in cost-effectiveness. Scatterplots of increments of costs versus increments of effectiveness emphasize the correlation between effectiveness and cost. This is driven by prolonged survival in costly health states such as TIA. The joint distribution is far from being elliptical (as would be implied, for example, by a normality assumption) and the correlation is stronger at lower efficiency values, as a result of an apparent ceiling on costs.

Summarization of this joint distribution requires care. A standard approach is to compute the ratio of the mean marginal effectiveness and the mean marginal cost. This is interesting but should be complemented by additional information. If a maximum acceptable MCER is given, as in the so-called league tables, one can calculate the probability that the cost-effectiveness of the intervention is smaller than this maximum, or ceiling. Graphing this

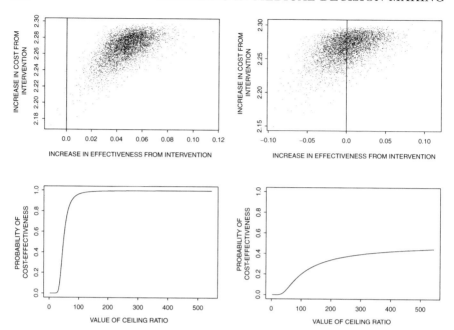

Figure 3.5 Top row: joint distribution of the increase in costs and increase in effectiveness from adopting the treatment, for individuals with stenosis (left) and without stenosis (right). Points to the right of the vertical bar have positive effectiveness. Bottom row: acceptability curves graphing the probability that the treatment is cost effective as a function of an MCER ceiling, expressed in dollars per unit of effectiveness. At the origin, the acceptability curve is the probability of a positive cost. As the ceiling approaches infinity, the acceptability curve approaches the probability of effectiveness.

probability as a function of the ceiling generates the so-called acceptability curve (van Hout *et al.*, 1994; O'Hagan *et al.*, 2000), illustrated in the bottom panels of Figure 3.5. For individuals with stenosis (left of Figure 3.5), the probability that the intervention will lead to improved life expectancy is virtually one. Then a probability interval can be a reasonable summary of the distribution of the MCER. For example, 98% of the sample points give MCERs between 28.66 and 150.42.

In this example the distribution of the MCER contains a small proportion of very large values, and is skewed, as we will see in more detail later. This arises because, in some draws, the effectiveness of intervention and that of standard care are very close. Thus, the denominator of the MCER is very small, implying that the MCER is large in absolute value. The mean of the distribution – that is, the mean of the ratios – can be a poor summary of the distribution of the MCER. The fact that the MCER distribution has a very long right tail is valuable information for policy making and

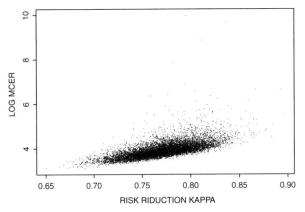

Figure 3.6   Scatterplot of samples from κ^1 and the logarithm of the MCER in patients with stenosis.

should not be ignored. Alternatively, more stable distributions are obtained by investigating marginal effectiveness/cost ratios rather than cost/effectiveness ratios.

More generally, summarization of the distribution of the MCER without an examination of the joint distribution of Figure 3.5 can lead to misleading analysis, when there are draws where the intervention reduces costs or results in a decrease in effectiveness (Heitjan *et al.*, 1999). In the first case the intervention dominates standard care, while in the second it is dominated. A more appropriate summarization is to estimate the probabilities of dominance and, separately, the MCER conditional on lack of dominance. Also, if both marginal cost and marginal effectiveness are near zero, the choice is a toss-up, as the strategies are nearly identical in terms of costs and benefits, but the MCER could be enormous. Inspection of a two-dimensional scatterplot such as Figure 3.5 would reveal this information.

In patients without stenosis (right of Figure 3.5), the distribution of the MCER includes a large portion of points for which the intervention is dominated by standard care. The probability that the intervention will lead to improved life expectancy is only slightly above 0.5. One-sided intervals are a practical solution in this case. For example, there is a probability of 28% that the intervention is not dominated and the MCER is less than 150.

Simulation-based probabilistic sensitivity analysis offers the opportunity to investigate in detail the relationship between the variability in individual inputs and the overall variability in the output. For example, Figure 3.6 illustrates the relationship between κ^1 and the MCER. A tiny fraction of simulated values with negative effectiveness, corresponding to the points to the left of zero in Figure 3.5, were eliminated. Because the distribution of effectiveness includes values very close to 0, the sampled MCER can assume very large values, and is graphed here on a log scale. Even on the log scale the distribution of the MCER is skewed towards larger values. As we would expect, variation in κ^1 accounts for a significant proportion of the overall variability in the MCER.

Representing the distribution of the MCER is again more complex for the cohort of patients without stenosis, where a large proportion of the sampled values have negative effectiveness. The top panel of Figure 3.7 considers the relationship between the risk reduction κ^0 and the MCER, using the subset of samples with positive effectiveness. The bottom panel graphs the probability of a positive MCER as a function of κ^0, estimated by the empirical frequencies over 25 bins, as well as the marginal distribution of κ^0.

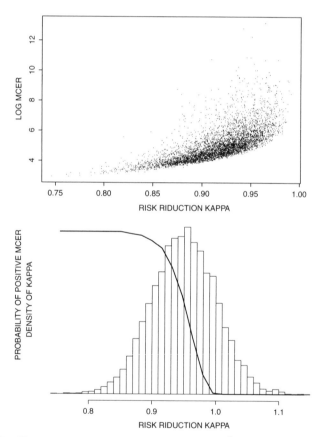

Figure 3.7 Top: scatterplot of samples from the κ^0 and the log of the MCER in patients without stenosis. Only positive values of the MCER are represented. Bottom: probability of a positive MCER as a function of κ^0, superimposed to a histogram of sampled values of κ^0. The probability curve ranges from 0 to 1.

3.5   Searching for strategies via simulation

In this section we discuss simulation approaches to searching for good strategies. These are useful when the number of actions to be considered is large and carrying out a complete evaluation of the model for each of them is impractical. We discuss some techniques in the context of an experimental design example, but the same approaches could be applied to other decision situations. A review of simulation-based utility analysis is offered by Müller (1998).

3.5.1   A two-phase design in screening for a rare disease

Estimating the prevalence of a rare disease by straightforward random sampling can require an impractically large sample before yielding reliable estimates. When possible, a more efficient strategy is to focus efforts on individuals that are likely to have the disease, for example using an inexpensive screening test based on symptoms, but doing so in a way that still allows a population-based estimate to be obtained. One implementation of this approach (Shrout and Newman, 1989; Erkanli *et al.*, 1998) is as follows. In the first phase of the study, the screening test is administered to a random sample of N subjects in the population. In the second phase, a proportion a_1 of the subjects who screened positively and a proportion a_2 of the subjects who screened negatively are administered a more accurate diagnostic test. The efficiency of this design depends critically on the choice of these proportions. Here we study the problem of finding N and proportions a_1 and a_2 which, subject to a budget constraint, give the maximum expected precision for estimating the unknown prevalence π.

We use the notation T+ (T−) for a positive (negative) screening result, and D+ (D−) for a positive (negative) diagnostic test. In this example we assume that the diagnostic test is fully accurate, so a positive (negative) diagnostic test coincides with the presence (absence) of the disease. Rather than parameterizing the statistical model in terms of sensitivity and specificity, as was done, for example, in Section 1.4, it is more convenient here to consider directly the positive and negative predictive values. Specifically, π^+ will be the probability of illness given a positive screening, that is, $\pi^+ = P(D+|T+)$; similarly, $\pi^- = P(D+|T-)$. Also, μ is the marginal probability of a positive screening test. The prevalence of disease is then $\pi = \mu\pi^+ + (1-\mu)\pi^-$.

The sampling distribution is simple to state. If we screen N subjects, a random y_0 of them will screen positively, with distribution $y_0 \sim \text{Bin}(N, \mu)$. If we apply the diagnostic test to proportions a_1 for positives and a_2 for negatives, the sizes of the two second-phase subsamples are $n_1 = a_1 y_0$ and $n_2 = a_2(N - y_0)$. Of these a random $y_1 \sim \text{Bin}(n_1, \pi^+)$ and $y_2 \sim \text{Bin}(n_2, \pi^-)$ will be found with disease in the two subgroups. The number of positive test results is collectively denoted by $y = (y_0, y_1, y_2)$.

We complete the model specification by choosing a prior distribution $p(\mu, \pi^+, \pi^-) = p(\mu)p(\pi^+)p(\pi^-)$, with $p(\mu) = \text{Beta}(q_0, b_0)$, $p(\pi^+) = \text{Beta}(q_1, b_1)$, and $p(\pi^-) = \text{Beta}(q_2, b_2)$. In our illustration we consider the case of childhood depression. Erkanli *et al.* (1998) developed choices of hyperparameters using reviews of the literature (Angold and Costello, 1993), and specified:

$$
\begin{array}{cccccc}
q_0 & q_1 & q_2 & b_0 & b_1 & b_2 \\
4.44 & 3.75 & 1.84 & 13.32 & 49.1 & 55.71
\end{array}
$$

An experimental design is characterized by N, a_1 and a_2, which jointly define an action. The cost and usefulness of an experiment cannot be known with certainty ahead of time, as they depend on the specific outcome y. The overall expected cost of an experiment is

$$C = Nc_s + (E\{\mu\}Na_1 + (1 - E\{\mu\})Na_2)c_d,$$

where $c_s = 1$ is the cost of a single screening test, and $c_d = 18$ the cost of a single diagnostic test. Again these values are based on Erkanli *et al.* (1998). The usefulness of the experiment can be measured in a variety of ways. Here we choose to capture it by the precision with which it allows us to estimate the prevalence a posteriori. Formally, the precision is given by the reciprocal of the posterior variance of π. This choice of utility leads to the same optimal design choice as the minimization of the squared error loss of estimation in standard Bayesian design setting (DeGroot, 1970), but we do not need to consider this explicitly here.

One approach to this problem is to explore the cost-effectiveness of alternative choices of N, a_1, and a_2. Instead, because of the large number of combinations, we approach it by setting the budget constraint that the overall expected cost should be $C = 2000$, and maximizing precision subject to this constraint. In particular, the constraint implies that

$$N = \frac{C}{c_s + (E\{M\}a_1 + (1 - E\{M\}))},$$

so that we only have two free decision parameters, a_1 and a_2. The utility can be expressed as

$$u(a_1, a_2, y) = \frac{1}{\text{Var}(\pi|y)},$$

and the goal is to maximize its expectation

$$\mathcal{U}(a_1, a_2) = \sum_{\text{every } y} \frac{1}{\text{Var}(\pi|y)} p(y) dy. \qquad (3.6)$$

We can obtain an explicit expression for $\mathrm{Var}(\pi|y)$ by noting that the a posteriori means and variances of μ, π^+ and π^- are respectively

$$m_i = \frac{q_i + y_i}{q_i + b_i + n_i},$$

$$v_i = \frac{(q_i + y_i)(b_i + n_i - y_i)}{(q_i + b_i + n_i)^2(q_i + b_i + n_i + 1)}$$

for $i = 0, 1, 2$ and $n_0 = N$. Then, as μ, π^+ and π^- are independent a posteriori,

$$\mathrm{Var}(\pi|y) = (v_0 + m_0^2)v_1 + (1 - 2m_0 + v_0 + m_0^2)v_2 + v_0(m_1 - m_2)^2.$$

In the remainder of this section we illustrate three simulation-based procedures for evaluating and maximizing (3.6): a straightforward Monte Carlo approach; a more efficient Monte Carlo approach that exploits the fact that $\mathcal{U}$ varies smoothly with the as; and an MCMC approach that samples as proportionally to their expected utility. In addition to these approaches, it is possible, although also computationally expensive, to evaluate (3.6) exactly. In more general problems, however, this is not always the case, and simulation-based approaches are the only practical choice.

Throughout, we emphasize that it is interesting to learn about many aspects of the expected utility surface (DeGroot, 1984), rather than just finding the optimal decision. There are several reasons for this, including gaining insight into the structure of the model, validating global optimality, and assessing changes in utility resulting from small deviations around the best decision (Kadane and Chuang, 1978). More generally, in practical applications, the utility function may not necessarily capture all relevant objectives (for design problems, see Box and Draper, 1975). It may then be convenient to use a utility function to model only the readily quantifiable objectives. By considering the entire surface, one can select a subset of good decisions, and, within that, choose the most satisfactory decision based on the criteria that are not as easily formalized. Ease of implementation, ease of communication of the results, and ethical concerns in clinical trials are common examples of such criteria.

3.5.2 Monte Carlo evaluation of the expected utility

A straightforward, and commonly used, approach for evaluating the expected utility surface $\mathcal{U}$ is to compute the value $\mathcal{U}$ point by point using a Monte Carlo approximation like (3.3), where the vector θ comprises both the unknown parameters and the unknown experimental results, and the function h represents the utility. This approach is easy to implement. The simulation is done in two steps: generating unknown parameters from the prior distribution; and generating unknown experimental outcomes from the likelihood function.

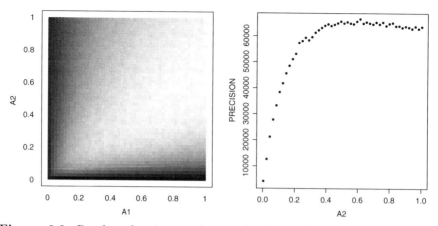

Figure 3.8 Results of pointwise large-scale Monte Carlo integration. Left: expected precision $\mathcal{U}$ of the estimate of prevalence as a function of design parameters a_1 and a_2. Darker gray shades correspond to lower expected utilities. The range of estimated values of $\mathcal{U}$ is $(3791.2, 67601.9)$. Right: a section of the surface for $a_1 = 0.89$.

It can be applied very broadly to complex experimental situations. The results of applying this approach to the problem we are considering are displayed in Figure 3.8, and will serve as the benchmark for evaluating two alternative approaches here and in Section 3.5.3. The Monte Carlo sample size in Figure 3.8 is 2000 at every point of a 50×50 grid, for a total of 5 million simulations and evaluations of u. Despite the massive scale of the simulation, the surface is still too rough to support reliable optimization, as emphasized by the section presented in the right-hand panel.

This consideration may motivate the implementation of a more efficient approach, which makes use of the fact that $\mathcal{U}$ is a smooth surface. When computing $\mathcal{U}(a)$, at a point $a = (a_1, a_2)$, it is efficient to borrow strength from the computation of $\mathcal{U}$ at neighboring values, which are likely to lead to similar expected utilities. This observation motivates replacing the sum over y in expression (3.6) by statistical curve fitting. If we let $\theta = (\mu_m, \pi_m^+, \pi_m^-)$ be the vector of model parameters, the expected utility surface can also be derived using the following algorithm:

1. Select a set of M design points $a_m = (a_{1m}, a_{2m})$, $m = 1, \ldots, M$, possibly including duplicate points.

2. For each a_m, simulate one point from the joint parameter and sample space (θ, y_m). This is accomplished by simulating parameters from the prior and then experimental results from the sampling distribution.

3. For each combination (a_m, θ_m, y_m) evaluate the utility $u(a_{1m}, a_{2m}, y_m)$, in our case the posterior precision.

4. Fit a smooth curve $\hat{\mathcal{U}}(a_1, a_2)$ through the points (a_{1m}, a_{2m}, y_m).

5. To maximize $\hat{\mathcal{U}}$, find deterministically the extreme point corresponding to the optimal design.

The curve fitting in step 4 is a regression, and therefore the resulting fit can be interpreted as an approximate conditional expectation of the utility given the action. Evaluation of the conditional expectation via curve fitting replaces evaluation via pointwise Monte Carlo integration. Depending on the application, fitting could be done by a parametric model or by any common scatterplot smoother such as loess or smoothing splines. We will refer to this approach as curve-fitting Monte Carlo (CFMC). Müller and Parmigiani (1995) and Müller (1998) provide additional discussion on the choice of smoothing algorithm, and on higher-dimensional implementations.

Figure 3.9 illustrates the CFMC algorithm applied to the design problem to hand. The panel on the right illustrates the mechanism of curve-fitting integration in two dimensions. The curve is based on the loess smoother (Chambers and Hastie, 1991). Gains in efficiency from CFMC over direct MC can be substantial. Figure 3.9 presents estimates of $\mathcal{U}$ based on CFMC, using 10 000 points, one per design choice over a 100×100 grid. The resulting curve is roughly as accurate pointwise as the large-scale Monte Carlo of Figure 3.8, but leads to far more stable optimization and is based on only 0.2% of the computing effort in terms of simulated values and utility evaluations.

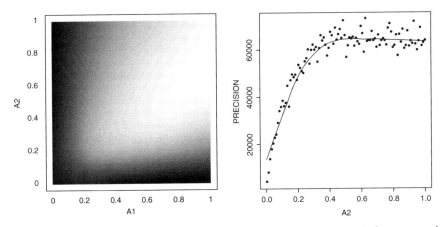

Figure 3.9 Results of curve-fitting Monte Carlo integration. Left: expected precision $\hat{\mathcal{U}}$ of the estimate of prevalence as a function of design parameters a_1 and a_2. Right: a section of the surface for $a_1 = 0.89$. The image on the left is based on smoothing of 10 000 realized values of the precision using loess. The image on the right is based on smoothing of 100 realized values of the precision, one per design point. These are shown as points.

3.5.3   Markov chain sampling of actions

An alternative approach for exploring expected utility surfaces is to draw samples of actions that have high expected utility. This is a method that is not generally as efficient as CFMC when they can both be implemented, but can be used to attack much more complex problems, for which no other approach exists. The general idea, proposed in Clyde *et al.* (1995) and generalized in Müller (1998), is to generate a sample of decisions by constructing a probability model on both the problem's unknowns and the decision variables. This can be done in such a way that the marginal distribution of the sampled actions is proportional to expected utility. Analyzing the sampled actions provides guidance in decision making. We will describe the algorithm in terms of a general nonnegative utility function $u(a, \theta, y)$, and then apply it to the design problem of Section 3.5.1, in which $a = (a_1, a_2)$ and the utility function is $u(a, \theta, y) = u(a_1, a_2, y) = 1/\text{Var}(\pi|y)$.

Consider the probability density function

$$h(a, \theta, y) \propto p_a(y, \theta) u(a, \theta, y). \tag{3.7}$$

The marginal distribution of a is

$$h(a) \propto \int h(a, \theta, y) d\theta dy = \mathcal{U}(a), \tag{3.8}$$

the expected utility of interest. The expected utility surface can be explored by studying samples of a, obtained using this augmented probability model. The optimal design is the mode of the marginal distribution (3.8). The marginal distributions of θ and y in the augmented model are no longer those of the original model and are not relevant for inference and prediction.

In some applications, the function $u(a, \theta, y)$ may be analytically integrated with respect to θ, giving

$$u(a, y) = \int_\Theta u(a, \theta, y) p(\theta|y) d\theta.$$

In other cases, such as our experimental design problem of Section 3.5.1, $u(a, y)$ can be specified directly. Then it is efficient to define the augmented probability model as

$$\bar{h}(a, y) \propto p_a(y) u(a, y). \tag{3.9}$$

Again, the marginal distribution of a is proportional to expected utility function,

$$h(a) \propto \int \bar{h}(a, y) dy = \mathcal{U}(a),$$

but variability is reduced. In order for (3.7) and (3.9) to define valid probability models, we need $h(a, \theta, y)$ and $\bar{h}(a, y)$ to be nonnegative and integrable. If necessary, integrability can often be guaranteed without practical consequences by constraining the action space to be a compact set.

To sample designs in proportion to their expected utility, we can set up an MCMC simulation scheme with $h(a, \theta, y)$ as the stationary distribution. The algorithm is summarized by the following steps:

1. Start with action a_0. Simulate (θ_0, y_0) from $p_{a_0}(\theta, y)$, by simulating θ from the prior distribution and then y conditional on θ, under a_0.

2. Generate a candidate $\tilde{a}$ from proposal distribution $g(\tilde{a}|a_0)$.

3. Simulate $(\tilde{\theta}, \tilde{y})$ as in step 1, under design $\tilde{a}$, generated in step 2.

4. Compute

$$\alpha = \min\left[1, \frac{h(\tilde{a}, \tilde{\theta}, \tilde{y})}{h(a_0, \theta_0, y_0)} \frac{g(a_0|\tilde{a})p_{a_0}(\theta_0, y_0)}{g(\tilde{a}|a_0)p_{\tilde{a}}(\tilde{\theta}, \tilde{y})}\right] = \min\left[1, \frac{u(\tilde{a}, \tilde{\theta}, \tilde{y})}{u(a_0, \theta_0, y_0)} \frac{g(a_0|\tilde{a})}{g(\tilde{a}|a_0)}\right].$$

5. Set
$$(a_1, \theta_1, y_1) = \begin{cases} (\tilde{a}, \tilde{\theta}, \tilde{y}) & \text{with probability } \alpha, \\ (a_0, \theta_0, y_0) & \text{with probability } 1 - \alpha. \end{cases}$$

6. Repeat steps 2–5 until the chain is judged to have converged to the desired level of accuracy.

This algorithm defines an MCMC scheme to simulate from $h(a, \theta, y)$, using a Metropolis–Hastings step to update (a, θ, y).

Specification of the proposal distribution $g(\tilde{a}|a)$ proceeds as discussed in Section 3.2.3. In particular, choosing a symmetric proposal distribution, that is, one for which $g(\tilde{a}|a) = g(a|\tilde{a})$, leads to the acceptance probability

$$\min\left[\frac{u(\tilde{a}, \tilde{\theta}, \tilde{y})}{u(a_0, \theta_0, y_0)}, 1\right].$$

In this case at each step the proposed action is accepted for sure if it is better than the current action, and it is accepted according to the ratio of utilities if it is worse.

A convenient choice for $g(\tilde{a}|a)$ is a normal distribution centered at a with covariance matrix Σ chosen to obtain acceptance rates between 25% and 75%. In a particular setup, Gelman et $al.$ (1996) discuss optimal rates and show when rates around 25% are optimal. The choice of proposal distribution does not affect the ergodic distribution of the chain, but only affects the speed of convergence. Alternatively, one can construct proposal distributions based on approximating the expected utility surface by a function $\hat{\mathcal{U}}(a)$ and taking $g(\tilde{a}|a) \propto \hat{\mathcal{U}}(a)$. In the terminology of posterior MCMC simulations, this would correspond to an independence chain step (Tierney, 1994).

Estimating the expected utility surface from the simulation output takes the form of a well-known problem in MCMC that is, reconstructing the marginal distribution of a subset of variables of interest. A straightforward estimate can be obtained using the empirical frequencies. Gelfand and Smith (1990) propose the use of 'Rao–Blackwellization' to reduce variation in the estimated density. To illustrate the concept, consider an additional step to draw $a \sim h(a|\theta, y)$ between steps 5 and 6. The Rao–Blackwellized estimate would take the form

$$\hat{h}(a) = \frac{1}{M} \sum_{m=1}^{M} h(a|\theta_m, y_m). \tag{3.10}$$

Here M is the Monte Carlo sample size, and (θ_m, y_m) is the value of (θ, y) at the mth iteration. Implementation of (3.10) requires in addition the evaluation of the normalizing constants of $h(a|\theta_m, y_m)$. These are

$$C_m = \left[\int_{\mathcal{A}} h(a, \theta_m, y_m) \mathrm{d}a \right]^{-1}.$$

Then (3.10) can be implemented as follows:

$$\hat{h}(a) = \frac{1}{M} \sum_{m=1}^{M} C_m h(a, \theta_m, y_m), \tag{3.11}$$

where (θ_m, y_m), $m = 1, \ldots, M$, are the simulated values of (θ, y) after M iterations of the Markov chain. The constants C_m can be computed by evaluating $\hat{h}(a)$ on a grid. If $\mathcal{U}$ is being evaluated on a fixed grid of points, evaluating the C_m involves no additional computing effort.

We can now compare the pointwise Monte Carlo results of Figure 3.8 with the output of the Markov chain. Figure 3.10 shows the estimated expected utility surface, based on the Markov chain before (left) and after Rao–Blackwellization (right). The gains from 'Rao–Blackwellizations' are substantial in this case. Yet the main features of the surface can be identified sufficiently well even from the raw frequency distribution on the right. The chain was run for 50 000 iterations, with the first 1000 iterations discarded as burn-in. Only every tenth iteration is used for the Rao–Blackwellization. Results stabilize early on; a much smaller chain would have been sufficient for a rough exploration of the surface. The average acceptance rate of proposals in the Metropolis–Hastings is 0.61.

The optimal design appears to be around $a^* = (1, 0.7)$, which would result in $N^* = 134$. As is typical of similar design problems, the expected utility surface is flatter near the maximum, and it is useful to consider regions of designs that have utility close to that of the optimum (Parmigiani and Berry, 1994). These are often called efficiency regions. In Figure 3.10, contours indicate isoefficiency curves at percentage levels 50, 75, 90, and 95, thus identifying the corresponding efficiency regions.

The Markov chain algorithm for sampling for utility functions applies to a wide class of problems. The only requirements are that for any action, and any

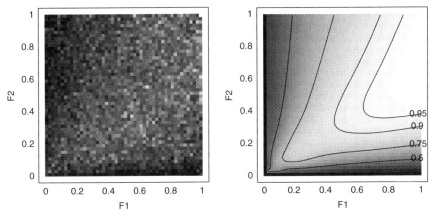

Figure 3.10   Precision of the estimate of prevalence as a function of design parameters a_1 and a_2. Empirical frequencies of the Markov chain for the example of Section 3.5.1, on a 50 × 50 grid. On the right, estimated expected utility surface for the same example, based on the Markov chain after 'Rao-Blackwellization'. Contours indicate isoefficiency curves at percentage levels 50, 75, 90, and 95..

values of the random variables (including unknown model parameters θ and observations y), the utility function can be evaluated; and that for any action, values of all random variables can be simulated from the model. In particular, the model does not have to allow for analytic posterior inference, or rely on restrictive choices of utility functions. While generally applicable, the Markov chain approach to design is likely to be more appropriate for problems in which evaluation of expected utility is difficult because of a complex structure in the probability model or because of a complicated definition of utility. The Markov chain algorithm can be inefficient if the utility surface is flat over a very large region (in which case identifying optima requires a very large sample size) or the action space is high-dimensional (in which case the chain is still useful for exploration, but estimation of the mode becomes prohibitive). Müller (1998) developed extensions that contribute to overcoming these problems.

Bielza *et al.* (1999) consider variations and extensions of the algorithm stated here to solving influence diagrams, including moderately sized sequential problems and possibly higher-dimensional decision parameters. Other stochastic schemes for choosing candidate decisions exist in other contexts. Examples include simulated annealing (Haines, 1987) and genetic algorithms. Both approaches require pointwise evaluation of the expected utility surface, while the approach discussed here does not.

Part II

Case Studies

4

Meta-analysis

4.1 Summary

In this chapter, we illustrate the use of Bayesian meta-analysis in developing probability distributions on the magnitude of the effects of a medical treatment. These distributions can be used informally to support decision making on patient treatment and on allocation of resources to future trials. They can also become components of a formal decision analysis, of a comprehensive decision model, or of a stochastic optimization. The chapter begins with a brief overview of meta-analysis, with pointers to the extensive, fascinating, and controversial literature. This is followed by an introductory example which looks at the efficacy of tamoxifen in adjuvant treatment of early breast cancer and serves as an introduction to some of the key features of Bayesian meta-analysis. The case study of this chapter deals with the synthesis of evidence from several clinical trials comparing the effectiveness of commonly recommended prophylactic treatments for migraine headaches. The case study is based closely on Dominici *et al.* (1999).

4.2 Meta-analysis

The term 'meta-analysis' originated in psychology. Glass (1976) used it to describe 'the statistical analysis of a large collection of results from individual studies for the purpose of integrating the findings'. In medicine, the practice of formally integrating findings from different studies can be traced back at least to the 1950s (Beecher, 1955). Today, meta-analysis has become a key component of evidence-based medicine. A MEDLINE search of articles published in 1997 yielded 775 articles with the term 'meta-analysis' in the title, abstract or keywords. This is the result of the growth in the number of clinical trials (approximately 8000 new clinical trials begin every year, according to Olkin 1995a) and of the desire to use accruing evidence as early as possible in improving health care decisions. Meta-analysis is also becoming widely applied beyond randomized clinical trials, for example in epidemiological research. Interesting discussions of the role of meta-analysis in clinical research and decision making will be found in L'Abbé *et al.* (1987) and Gelber and Goldhirsch (1991).

Moher and Olkin (1995) summarize the reasons for the success of meta-analysis:

> Why the dramatic increase in the number of published meta-analyses? Two examples may help address the question. A meta-analysis published in 1990 described the efficacy of corticosteroids given to mothers expected to deliver prematurely (Crowley *et al.*, 1990). The results of the meta-analysis indicated that corticosteroids significantly reduced morbidity and mortality of these infants. This analysis convincingly showed that such evidence was available at least a decade earlier ([i.e.] 1980). Had a meta-analysis been conducted when the evidence became available, much unnecessary suffering might have been avoided.
>
> In an article on meta-analyses published in *JAMA* in 1992 (Antman *et al.*, 1992), the research team showed that textbook recommendations for the treatment of patients with suspected myocardial infarction often lagged behind the empirical evidence by as much as 10 years. The group also noted that, at times, the opinion of experts writing these books was in sharp contrast to the empirical evidence.

In general, meta-analyses can offer important advantages over more traditional narrative approaches to the overview of scientific evidence (Chambers and Altman, 1994). These advantages result primarily from the systematic, explicit, and quantitative nature of the synthesis provided by meta-analysis; from the possibility of assessing uncertainty about the results of the synthesis; and from the increase in sample sizes deriving from the combination of studies. From the point of view of the philosophy of science, meta-analysis is a novel paradigm for scientific investigation, reflecting the scientific community's adaptation to the information explosion of the last few decades. Goodman (1998) hails meta-analysis as 'one of the most important and controversial developments in the history of science'.

As with many promising new paradigms, meta-analysis has given rise to misuses and controversy. Criticisms of meta-analysis are both conceptual and methodological. Randomized clinical trials are modeled on controlled experimentation and are generally agreed to address in a satisfactory way the question of the causal relationship between treatment and outcomes. Difficulties may arise: the sample size can be small, the measurable outcomes may problematic, the protocol may be difficult to implement. But the paradigm is thought to be sound. The design of a meta-analysis study is different from that of a randomized clinical trial. Some authors (such as Erwin, 1984) consider these differences to be sufficient to question the possibility of inferring causal relations from meta-analysis of clinical trials (even though it is possible to infer such relations from the individual trials). A related point is made by Charlton (1996), who writes that 'the prestige of meta-analysis is based upon a false model of scientific practice'; meta-analysis, in his view, cannot be considered a hypothesis testing activity, and should be confined to effect size estimation.

Another common criticism is epitomized by the slogan 'many bad studies don't make a good one'. The argument is this: for a given clinical question there either is a single, critical, well-conducted, study, with a large enough sample, that can provide guidance to physicians, or there is not. If there is not, that is often because existing trials are conflicting, diverse, or not sufficiently well conducted. Meta-analysis, it is argued, should not be used to attempt to settle a clinical question in the presence of vexing primary data problems.

Methodologically, critical issues are search strategies, publication bias, and study heterogeneity. Some meta-analyses are conducted by gathering primary data from the study investigators (Simes, 1986; Early Breast Cancer Trialists Collaborative Group (EBCTCG), 1990). This is an appropriate and effective strategy, but it is not always feasible. Most meta-analyses are based on published results. The key elements of a meta-analysis of published results are the criteria used for searching for, evaluating, and selecting articles. Concerns about the quality of these criteria in applied research are common (Cook et al., 1995; Sacks et al., 1987; Chambers and Altman, 1994). Guidelines for rigorous procedural methodology have been put forward by several authors (Simes, 1986; Deeks et al., 1997) and by ad hoc working groups in clinical trials (Moher et al., 1999) and epidemiology (Stroup et al., 2000).

Publication bias arises because scientific journals selectively publish studies with statistically significant results. For example, in 1986–87 about 76% of the articles published in the New England Journal of Medicine used statistical tests; 88% of these tests rejected the null hypothesis. These proportions are even higher in the experimental psychology literature (Sterling et al., 1995). It is clear that the results of published studies are not a representative sample of the results of all studies, and that this may systematically bias the result of a meta-analysis which considers only published results; for example, a small treatment effect is likely to become artificially magnified. Diagnostics for the presence of publication bias are based on observing a relationship between sample size and effect size (Duval and Tweedie, 2000).

Because the aims of a meta-analysis are typically broader that those of the individual trials being reviewed, it is likely that any sizable meta-analysis will have several sources of between-trial heterogeneity. These include differences in specific treatment regimens, patient eligibility criteria, baseline disease severity, and outcomes. Complete homogeneity is not necessarily a desirable goal. Moher and Olkin (1995) note that 'sometimes, too much homogeneity of studies will stifle generalizations to a larger population. On the other hand, too much study heterogeneity will weaken the results. Thus, there has to be an understanding of the sources of the heterogeneity.' Similar views are expressed by Thompson (1994): 'discussion of heterogeneity in meta-analysis affects whether it is reasonable to believe in one overall estimate that applies to all the studies encompassed, implied by the so called fixed effect method of statistical analysis. Undue reliance may have been put on this approach in the past, causing overly simplistic and overly dogmatic interpretation.' Statistical models and techniques for quantifying heterogeneity and for developing and interpreting summary estimates are illustrated extensively in the rest of this chapter.

There are also issues surrounding the quality of reporting of meta-analyses. Jadad and McQuay (1996) carried out a systematic review of methodology in 74 meta-analyses of analgesic interventions. They found that: 'Ninety percent of the meta-analyses had methodological flaws that could limit their validity. The main deficiencies were lack of information on methods to retrieve and to assess the validity of primary studies and lack of data on the design of the primary studies'. They also found that 'meta-analyses of low quality produced significantly more positive conclusions'. Sacks *et al.* (1987) identify six content areas thought to be important in the conduct and reporting of meta-analyses: study design, combinability, control of bias, statistical analysis, sensitivity analysis, and problems of applicability. Moher and Olkin (1995) review the issues and lay the groundwork for developing standards for the reporting of meta-analyses.

In this chapter we will discuss some statistical tools for meta-analysis. What is their role amidst this controversy? I will take a pragmatic view: there are decisions to be made today, and we should make them using the best available evidence. The limitations of mathematical modeling as a means of synthesis can be serious, but a well-conducted analysis can often limit publication bias, incorporate heterogeneity, and provide practical guidance.

4.3 Bayesian meta-analysis

Statistical methods have been developed for meta-analysis and are continuously being refined in response to the increasing demand for meta-analysis and the increasing complexity of the meta-analyses performed. Olkin (1995b) provides a historical perspective; Sutton *et al.* (1998) provide a comprehensive and up-to-date review; Hasselblad and McCrory (1995) give a more concise practical guide; Stangl and Berry (2000) provide a collection of state-of-the-art applications. A classic and beautiful book on the subject is that by Hedges and Olkin (1985). Reviews of software include Sutton *et al.* (2000) and Normand (1995).

Here we will be concerned with Bayesian methods, whose use is well established in the statistical literature (DuMouchel and Harris, 1983; Berry, 1990; DuMouchel, 1990; Eddy *et al.*, 1990) and is gaining acceptance in the medical literature as well (Baraff *et al.*, 1993; Berry, 1998; Biggerstaff *et al.*, 1994; Brophy and Joseph, 1995; Sorenson and Pace, 1992; Tweedie *et al.*, 1996). Many interesting situations can now be modeled using software packages such as BUGS. Detailed applications of meta-analysis are illustrated by Smith *et al.* (1995; 2000).

Interest in Bayesian meta-analysis is motivated by several desiderata:

a) providing decision makers with summaries of evidence in the form of probability distributions, given all available evidence. This input is appropriate for a subsequent decision model. For example, Pallay and Berry (1999) demonstrate this using Bayesian meta-analysis to assess the worthiness of a phase III trial.

b) developing approaches for modeling trial heterogeneity, and devising summary measures that are relevant to decision making in the presence of trial heterogeneity. For example, it is important to address the question of the probability that a patient receiving drug A survives longer than a patient receiving drug B for a patient from a future, or unobserved, or hypothetical trial. This is a question of prediction. Bayesian random effects and hierarchical models (Lindley and Smith, 1972; Raudenbush and Bryk, 1985; Morris and Normand, 1992; Carlin and Louis, 2000) provide a flexible and practical framework for developing predictive models.

c) modeling unobserved aspects of the data generation and reporting processes. Examples include modeling publication bias (Silliman, 1997; Givens *et al.*, 1997), missing covariates (Lambert *et al.*, 1997), and partially reported results (Dominici *et al.*, 1999).

d) realistically assessing uncertainty. Bayesian simulation-based methods do not need to rely on asymptotic approximations and can straightforwardly accommodate uncertainty about nuisance parameters, often leading to more conservative and accurate statements about uncertainty in the overall conclusions. For example, Carlin (1992) finds that, in the meta-analysis of 2×2 tables, Bayesian estimates of parameter uncertainty are more accurate than the corresponding empirical Bayes estimates.

The limitations of Bayesian meta-analysis are related primarily to the added complexity of implementation. Depending on the application and the state-of-the-art in the field, elicitation of prior information may also become complex or controversial.

4.4   Tamoxifen in early breast cancer

4.4.1   Background

To illustrate some of the interesting features of Bayesian meta-analysis in a simple and common situation, let us consider a case in which each study reports a 2×2 table of successes and failures for both a treatment and a control group. This situation is exemplified by the data of Table 4.1, taken

MODELING IN MEDICAL DECISION MAKING

Table 4.1 Summary results for the tamoxifen and control group in 14 randomized clinical trials, as reported by EBCTCG (1998a). Breast cancer recurrence rates are abbreviated as Rec.

	Tamoxifen			Control			
Study s	Rec. x_1^s	Total n_1^s	Odds of of rec. $\frac{x_1^s}{n_1^s-x_1^s}$	Rec. x_0^s	Total n_0^s	Odds of rec. $\frac{x_0^s}{n_0^s-x_0^s}$	Odds ratios $\frac{x_1^s/(n_1^s-x_1^s)}{x_0^s/(n_0^s-x_0^s)}$
1	55	97	1.310	67	101	1.971	0.665
2	137	282	0.945	187	306	1.571	0.601
3	505	927	1.197	590	915	1.815	0.659
4	62	123	1.016	74	140	1.121	0.907
5	99	239	0.707	118	236	1.000	0.707
6	50	130	0.625	49	107	0.845	0.740
7	185	311	1.468	200	319	1.681	0.874
8	186	303	1.590	187	307	1.558	1.020
9	148	325	0.836	178	325	1.211	0.691
10	25	79	0.463	38	86	0.792	0.585
11	223	344	1.843	224	350	1.778	1.037
12	183	937	0.243	185	936	0.246	0.985
13	2	12	0.200	0	8	0.000	
14	129	434	0.423	159	449	0.548	0.771

from the overview of clinical trials of adjuvant tamoxifen for women with early breast cancer carried out by the EBCTCG (1998a). Since the mid-1980s this group has been responsible for thorough and influential reviews of randomized clinical trials of all treatments for early breast cancer. Their analysis is based on patient-level data obtained from study investigators, rather than on published summaries. As a result it is likely to be very robust to publication bias. Extensive effort has been devoted to the grouping of trial results according to relevant clinical characteristics, such as type and duration of treatment, dose of drug, use of other therapies in conjunction with tamoxifen, patient prognostic factors, and type of outcome (recurrence or death). Data collection and data checking procedures are described in detail in EBCTCG (1990; 1998a).

The data of Table 4.1 refer to 14 randomized clinical trials, as reported by EBCTCG (1998a). Each trial compared a group receiving tamoxifen for an average of about one year with a control group not receiving tamoxifen. We consider the endpoint to be breast cancer recurrence. There is variation in odds of recurrence among trials. In this case, this variation is mainly attributable to different follow-up times at the time of the overview, and different patient selection criteria. The ratios of the odds of recurrence in the two arms also vary (see also Figure 4.4 below). Modeling this variation is the main focus of this section.

4.4.2 Modeling heterogeneity

One approach to analyzing several 2×2 tables is to first perform a preliminary test for heterogeneity. If the null hypothesis (of homogeneity) is not rejected, it is common to proceed with analyses that effectively pool patients as though they all belonged to the same trial. If the null hypothesis of homogeneity is rejected, it is common to declare the studies too dissimilar to be combined, and stop. This general two-step approach can be implemented using both Bayesian and frequentist tools. For example, classical tests for heterogeneity are reviewed by Sutton *et al.* (1998), while a Bayesian approach using Bayes factors is proposed by Berry (1999).

When the meta-analysis is carried out to support decision making, this approach has several limitations. One problem is that, unless the number of trials is very large, tests have insufficient power to detect heterogeneity. Also, when the trials are sufficiently similar in clinical design, a moderate amount of heterogeneity can be desirable, because it can give a sense of how well the conclusion of trials can be extrapolated to other clinical settings and populations. The relevant quantities for clinical decisions refer to predictions for future patients. How will these be affected by the choice of a homogeneity versus a heterogeneity analysis? If heterogeneity is indeed small, the results will not deviate much. But the larger the heterogeneity, the more important it is to acknowledge it. So in an analysis whose goal is to assist clinical decision making, the scientific modeling issue is not so much whether the trials are homogeneous or not. In the vast majority of meta-analyses they are not. The practical issue is whether there is enough *prima facie* evidence of heterogeneity to justify the small additional trouble of implementing a heterogeneity model.

A step toward quantifying the heterogeneity of trials, and incorporating this into decision making if necessary, is to use a model with two levels: one describes the population of trials; the other describes the subpopulations of patients within each trial. These multilevel models are usually called hierarchical, because of the nesting of patients within trials (see also Figure 4.1 below). To set up a two-level hierarchical model for the tamoxifen example, we begin with some notation. Studies will be indexed by s. The sample sizes in the control and treatment arms of study s are n_0^s and n_1^s, respectively. The numbers of observed recurrences x_0^s and x_1^s. All trials are randomized, and therefore the sample sizes carry no information about efficacy. In study s, the probability distribution of each 2×2 table is completely characterized by two parameters: the probability of recurrence in the control arm (π_0^s) and in the treatment arm (π_1^s).

It is useful to reparameterize these into a measure of baseline recurrence rate (π_0^s) and a measure of the efficacy of treatment. Two measures of the efficacy of treatment are the log odds ratios

$$\lambda^s = \log\left(\frac{\pi_1^s}{1 - \pi_1^s}\right) - \log\left(\frac{\pi_0^s}{1 - \pi_0^s}\right), \qquad s = 1, \ldots, 14,$$

and the log relative risks

$$\log\left(\frac{\pi_1^s}{\pi_0^s}\right), \qquad s = 1, \ldots, 14;$$

both of these become smaller the more effective the treatment. In our discussion we will model the log odds ratios λ^s. Approaches based on modeling the log relative risks can be carried out along the same lines and require only small modifications to the procedure. One additional difficulty is that for any given baseline recurrence rate π_0^s, the range of values for the relative risk is constrained: the relative risk cannot be greater than $1/\pi_0^s$. In parametric models, the choice of parameterization can be important, especially when the number of studies is not large. One choice criterion is interpretability. For example, it may be more natural to specify distributional assumptions or to elicit expert opinion on one parameterization.

Let us consider the (π_0^s, λ^s) parameterization. The log odds ratios λ^s are likely to vary from study to study, but on the other hand are also likely to reflect the actual underlying efficacy of tamoxifen. Knowing the value of the log odds ratios in one study is informative about log odds ratios in other studies, despite heterogeneity. Assuming that all log odds ratio are the same (as assumed by the fixed effects models) ignores heterogeneity. Assuming that they are independent would ignore a key commonality. We need a compromise between independence and equality.

One way to achieve this compromise is to postulate a hypothetical population of studies, each with a different log odds ratio λ^s, and then learn about the features of this population from the data. We will discuss the simple case $\lambda^s \sim N(\theta, \tau^2)$. Here θ represents the overall mean log odds ratio and plays a role similar to the common log odds ratio in a model without heterogeneity; τ measures the study-to-study variability in the log odds ratios. Both θ and τ will be unknown and inferred from the data. In this way, it is the observed data that help us decide where we should position our analysis in the continuum between independence and equality of the effects. Alternative distributional assumptions are also possible. For example, assuming that the λ^s have a Student t or a double exponential distribution would result in inferences that are more robust to unusually large or small effects.

In the tamoxifen example, knowing the baseline rate π_0^s of one study is unlikely to inform us about the baseline rate of other studies. The determinants of the baseline rates are likely to be more strongly related to study design and current follow-up than to underlying clinical factors. Therefore, it can be reasonable to assume that the π_0^s, unlike the λ^s, are independent of each other, and forgo modeling a second-stage distribution. In other examples it may be appropriate to model a second-stage distributions for both the λ^s and the π_0^s.

In this analysis we are assuming that studies are conditionally independent given θ and τ. A consequence of this is that we are assuming that the studies are exchangeable, or in other words that that there are no study features that can help us predict whether the log odds ratios in a study is more or less likely to be, say, larger than average. A challenge to this assumption comes from the possibility that the log odds ratios might be correlated with baseline rates across studies. A simple diagnostic for this assumption is to plot empirical log odds ratios versus empirical recurrence probabilities in the control arms, and look for relationships. For example, if the treatment began to take effect only after a certain number of months, we could observe a positive relationship, with trials having longer follow-up displaying both a higher mortality and larger effect. In our case, the plot reveals no relationship.

Figure 4.1 summarizes the interconnections among the variables in the model. All variables are probabilistically dependent. However, the model is specified via conditional independence assumptions that simplify model-building, interpretation and computing. For example, the data in study 1 are conditionally independent of θ and τ given λ^1. However, via λ^1, the data in study 1 do provide information about θ and τ. In general, unknown parameters can be thought of as vehicles for information to percolate across the model. If the value of λ^1 were revealed to us by an oracle, the data in study 1 would not provide any information about θ and τ.

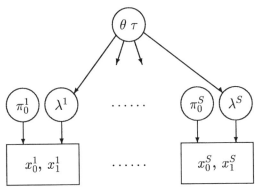

Figure 4.1 Graphical summary of the relationships among the parameters and observations in the hierarchical model considered in this section. Circles represent unknown parameters; rectangles represent data. Links represent probabilistic dependence relationships. The parameters τ and θ describe the population of study-specific log odds ratios. Each study also has a baseline recurrence rate π_0^s, but in this model these are independent of each other. This is reflected in the fact that they are not linked by belonging to a higher-level population.

The Bayesian model is completed by prior distributions on the π_0^s, θ, and τ. Our overall strategy here is to specify a vague prior distribution on the π_0^s and θ, which are well estimated based on the data, and a more informative prior distribution for τ. All priors are proper, in that they yield 1 when integrated over all possible parameter values. In the case of τ, it is important to specify a proper prior, as familiar choices of noninformative improper priors may lead to improper posterior distributions (DuMouchel, 1990).

More specifically, we assume that the π_0^s are independent and uniformly distributed over $(0, 1)$, and that θ follows a normal distribution with large standard deviation (say, 10) and mean zero. Because of the large standard deviation, the choice of the prior mean is unimportant. Because τ reflects sources of variability that are likely to be common to other groups of similarly heterogeneous trials, a possible strategy is to resort to other meta-analyses to gather a priori information about the likely magnitude of τ. Smith *et al.* (2000) illustrate this technique using a large collection of unrelated meta-analyses. In our case, because the focus is solely on the trials of Table 4.1, we could consider two other groups of trials of tamoxifen for the same type of patients, where tamoxifen was administered for two and five years instead of one. One caveat is that the heterogeneity may increase with the magnitude of the effect, so that heterogeneity of trials with longer treatment may be slightly higher.

A convenient choice of functional form for the distribution of τ^2 is to specify an inverse gamma distribution. We can use the parameterization with density function $f(x) = b^a x^{-a-1} e^{-b/x} / \Gamma(a)$ for $x > 0$. The mean is $b/(a - 1)$ and the variance $b^2/(a - 1)^2 (a - 2)$. We selected hyperparameters $a = 3$, the integer value giving the most diffuse finite-variance distribution, and $b = 0.1$, to approximately match the dispersion from the other groups of trials. One way of interpreting this prior specification in the simpler, nonhierarchical, problem is that it roughly corresponds to the information provided by $a = 3$ previous observations whose sum of squared deviations from the mean is $b = 0.1$. Specifying $a = 3$ limits the sensitivity of the analysis to the specified value of b, which is likely to be overridden by the experimental information. Because reasonable choices of prior distributions for τ can lead to different results, a sensitivity analysis, comparing the results obtained under different priors, is generally useful. While we do not illustrate it in this section, we will return to this point in Section 4.6.

In summary, we specified the following model:

$$
\begin{array}{ll}
\text{PRIOR} & \begin{aligned} \theta &\sim N(0, 100) \\ \tau^2 &\sim \mathrm{IG}(3, 0.1) \end{aligned} \\[2ex]
\text{STUDIES} & \begin{aligned} \pi_0^s &\sim U(0, 1) \\ \lambda^s | \theta, \tau^2 &\sim N(\theta, \tau^2) \end{aligned} \quad s = 1, \ldots, S \\[2ex]
\text{PATIENTS} & \begin{aligned} x_0^s | \pi_0^s &\sim \mathrm{Bin}(\pi_0^s, n_0^s) \\ x_1^s | \pi_1^s &\sim \mathrm{Bin}(\pi_1^s, n_1^s) \end{aligned}
\end{array}
$$

There are two parameters of primary interest. The first is the mean log odds ratio θ in the population of trials. Inference about θ addresses the question of the size of is the effect shown by the trials. The second is the log odds ratio λ^{S+1} in a future, unobserved, or hypothetical trial based on the available population of trials. The latter predictive distribution is directly relevant for clinical decisions. In particular, a possible future trial is a single woman who is faced with the decision about tamoxifen as an adjuvant therapy for early breast cancer. The benefit she will receive can be regarded as the next observation from the population of trial benefits.

Models for combining 2×2 tables in meta-analysis while acknowledging heterogeneity have been proposed by several authors. A seminal paper in this area is DerSimonian and Laird (1986). There are also Bayesian hierarchical models that share a similar overall structure and goals with the model of this section. Carlin (1992) and Smith *et al.* (2000) reparameterize the π^s into the corresponding logits, and assign normal conjugate distributions to them; Berry (1998) considers a Poisson, rather than binomial, sampling scheme, and models the event rates (or hazard rates) in the control group using a gamma with unknown parameters, and the log of the ratio of hazard rates in the two groups by a normal. A significant difference between the strategy used here and all these is that while the efficacy parameters (log odds ratios) are modeled hierarchically, here the baseline event rates are not.

4.4.3 Computing

To evaluate the posterior and predictive distributions we will use an MCMC algorithm, sampling in turn from τ, θ, and the pairs (π_0^s, λ^s). The full conditional distributions are

$$\theta \sim N\left(\left(\frac{1}{\tau^2}\sum_s \lambda^s\right)\left(\frac{1}{100}+\frac{S}{\tau^2}\right)^{-1}, \left(\frac{1}{100}+\frac{S}{\tau^2}\right)^{-1}\right)$$

$$\tau^2 \sim \text{IG}\left(3+\frac{S}{2}, 0.1+\frac{1}{2}\sum_s(\lambda^s-\theta)^2\right)$$

$$(\pi_0^s, \lambda^s) \sim K(\pi_0^s)^{x_0^s}(1-\pi_0^s)^{n_0^s-x_0^s}\left(1+\frac{1-\pi_0^s}{\pi_0^s}e^{-\lambda^s}\right)^{-n_1^s}\left(\frac{1-\pi_0^s}{\pi_0^s}e^{-\lambda^s}\right)^{n_1^s-x_1^s}$$
$$\times e^{-\frac{1}{2\tau^2}(\lambda^s-\theta)^2},$$

where K is the normalizing constant.

The first two full conditional distributions are the same as the posterior distributions for the mean of a normal population with known variance and for the variance of a normal population with known mean. These can be sampled directly. General expressions which do not depend on the specific hyperparameters chosen here, and details of the derivations, can be found in (Bernardo and Smith, 1994).

The full conditional distributions of the recurrence probabilities are not immediately recognizable. To sample from them we can use a Metropolis–Hastings step within the MCMC method (Tierney, 1994). There are many ways to implement this step. We used the strategy of approximating the conditional covariance matrix of (π_0^s, λ^s) and using it in a symmetric random walk Metropolis algorithm (see Section 3.2.3). First, we approximated the binomial component of the full conditional (the top line) with a bivariate normal. Using the central limit theorem, the statistics x_t^s/n_t^s are independent with limiting distributions

$$\sqrt{n_t^s}\left(\frac{x_t^s}{n_t^s} - \pi_t^s\right) \to N(0, \pi_t^s(1 - \pi_t^s)), \qquad t = 0, 1.$$

To simplify the presentation we denote this by

$$\frac{x_t^s}{n_t^s} \sim N\left(\pi_t^s, \frac{\pi_t^s(1 - \pi_t^s)}{n_t^s}\right), \qquad t = 0, 1.$$

Using the multivariate Delta method (described in a similar application in (Bishop *et al.*, 1975, pp. 486ff.), we can transform the implied joint distribution into an approximate distribution for $x_0^s/n_0^s, \hat{\lambda}^s$, where $\hat{\lambda}^s = \log(x_1^s/(n_1^s - x_1^s)) - \log(x_0^s/(n_0^s - x_0^s))$ is the maximum likelihood estimate of the log odds ratio λ^s based on the study-specific information alone. This asymptotic distribution is

$$\begin{bmatrix} \frac{x_t^s}{n_t^s} \\ \hat{\lambda}^s \end{bmatrix} \sim N\left(\begin{bmatrix} \pi_t^s \\ \lambda^s \end{bmatrix}, \begin{bmatrix} \frac{\pi_0^s(1-\pi_0^s)}{n_0^s} & -\frac{1}{n_0^s} \\ -\frac{1}{n_0^s} & \frac{1}{\pi_0^s n_0^s} + \frac{1}{(1-\pi_0^s)n_0^s} + \frac{1}{\pi_1^s n_1^s} + \frac{1}{(1-\pi_1^s)n_1^s} \end{bmatrix}\right) \equiv N(m_0, V_0).$$

Substituting this approximate normal likelihood for the binomial terms into the full conditional distribution of (π_0^s, λ^s) leads to the product of two normal terms. Replacing V_0 with a point estimate $\hat{V}_0$ and combining the quadratic forms on the exponents of these terms, we obtain a normal approximation to the full conditional. The variance of this approximation is

$$V_1 = \left(\hat{V}_0^{-1} + \begin{bmatrix} 0 & 0 \\ 0 & \frac{1}{\tau^2} \end{bmatrix}\right)^{-1},$$

while the mean is

$$m_1 = \hat{V}_0^{-1} m_0 + \begin{bmatrix} 0 \\ \frac{\theta}{\tau^2} \end{bmatrix}.$$

In general, $\hat{V}_0$ can be estimated by the study-specific maximum likelihood, that is, by substituting empirical frequencies for recurrence probabilities. In this example, however, $\hat{V}_0$ was obtained by adding 1 to each count before computing the empirical frequencies. In this way, probabilities on each arm

```
model;
{
    for( i in 1 : Num ) {
        x0[i] ~ dbin(p0[i],n0[i]) ;
        x1[i] ~ dbin(p1[i],n1[i]) ;
        p0[i] ~ dunif(0,1);
        logit(p1[i]) <- logit(p0[i])+lambda[i] ;
        lambda[i] ~ dnorm(theta,tau);
    }
    theta ~ dnorm(0.0, 0.01);
    tau ~ dgamma(3, 0.1);
}
```

Figure 4.2 BUGS command-line code for the model specification in Section 4.4.2.

are approximated by $(x + 1)/(n + 2)$. This is a somewhat *ad hoc* procedure, but it is useful for avoiding infinite variance estimates for the log odds when no recurrences are observed. There is a Bayesian interpretation for it: $(x+1)/(n+2)$ is the posterior mean in a single-study analysis when assuming a uniform prior distribution on the recurrence probabilities.

This approximation to the full conditional distribution provides an excellent fit for studies with larger sample sizes, but is less accurate for small studies, and turns out to be quite poor in study 13, which only has 8 and 10 patients in the two arms. If all studies were large, we could use this distribution to develop an independent proposal Metropolis–Hastings sampler, or even use the approximation directly in lieu of the actual full conditional. In this data set, however, it is safer to implement a symmetric random walk proposal, in which a proposed value is obtained by perturbing the current value. Our approximation is still crucial to determining the covariance matrix of the perturbation.

When utilizing a random walk proposal, we have the flexibility of using a variety of symmetric distributions to generate a new parameter value. In this case, two natural choices are the normal and Student's t. We used the latter here, to allow for better sampling of the tails of the full conditional distributions in studies with small sample sizes. Student's t is centered at the current parameter value, has scale matrix dV_1, and degrees of freedom given by the smallest sample size in all study arms (in our case 8). d is a scale factor that needs to be adjusted based on the application to hand. Values between 1 and 4 should provide good performance and are unlikely to need adjustment.

The model considered here, as well as many other useful models for combining 2×2 tables can be fit using the software package BUGS, mentioned in Section 3.2.3. Figures 4.2 and 4.3 show how to enter the data and how to

```
list( x1 = c( 55, 137, 505,  62,  99,  50, 185,
             186, 148,  25, 223, 183,   2, 129),
      n1 = c( 55, 137, 505,  62,  99,  50, 185,
             186, 148,  25, 223, 183,   2, 129),
      x0 = c( 67, 187, 590,  74, 118,  49, 200,
             187, 178,  38, 224, 185,   0, 159),
      n0 = c(101, 306, 915, 140, 236, 107, 319,
             307, 325,  86, 350, 936,   8, 449),
      Num=14)

list(lambda = c(0,0,0,0,0,0,0,0,0,0,0) ,
     tau = 1,
     theta = 0,
     p0 = c(0.5,0.5,0.5,0.5,0.5,0.5,0.5,0.5,0.5,0.5,0.5))
```

Figure 4.3 BUGS command-line code for the dataset of Table 4.1, to be used in conjunction with the model specification in Figure 4.2.

represent the model of this section. In this framework it is also straightforward to explore the sensitivity of results to alternative choices of distributions for the λs. BUGS can handle, for example, the double exponential and Student's t.

4.4.4 Results

Results are summarized in Figure 4.4 in terms of study-specific relative risks. A similar figure in terms of log odds ratios can be produced using the same MCMC output. Horizontal bars represent 98% probability intervals. Their width reflects the accuracy with which relative risks can be estimated within each study. For example, even though studies 3 and 12 have approximately the same sample size, the posterior probability distributions of the study-specific relative risks have very different intervals. This is the result of their different baseline event rates: study 3 has an event rate closer to $1/2$. Studies with wider intervals contribute less to the overall conclusions. One way to see this is to realize that their relative risks will fluctuate more widely in the simulation.

Within each study, the relative risk is π_1^s/π_0^s. For the next trial we can derive the predictive distribution of the log odds ratios directly from the model. We then convert this distribution into a distribution for the relative risk by fixing the baseline rate (in this example to $1/2$) and performing a change of variables. In practical situations, the baseline rates of an adverse event may be known based on patient prognostic factors. Then the predictive distribution of the absolute recurrence rate π_1^s can be derived from the baseline rates and the distribution of log odds ratios or relative risks. If the baseline rate is not known exactly, but a probability distribution is available, the same analysis applies with straightforward modifications.

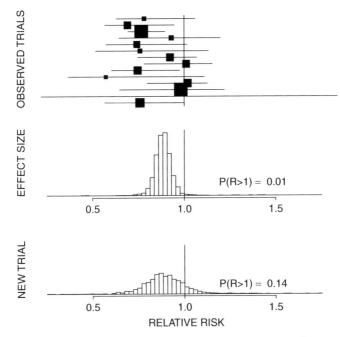

Figure 4.4 Summary of meta-analysis results for the tamoxifen example. The horizontal scale is relative risk, that is, the ratio of recurrence rates in the treatment to the control group. Values smaller than one correspond to a beneficial treatment. Squares in the top panel represent empirical relative risks. Their area is proportional to the overall sample size in the trial. Horizontal lines cover 98% of the probability distributions of π_1^s/π_0^s, with 1% of the mass removed on either side. The interval for study 13 is wider than the chosen limits of the figure. The middle panel is the posterior distribution of the overall relative risk, assuming a baseline recurrence rate of 50%. The bottom panel is the predictive distribution of the relative risk in a future trial, assuming again a baseline recurrence rate of 50%. The probability that the average effect in a trial is less than one is less than 0.99, but the probability of a benefit in a future patient is 0.86. Both probabilities are independent of the assumed baseline rate.

It is interesting to contrast predictions for a future patient with inferences about the average effect size. The probability of a benefit in a future patient is 0.84. (This is independent of the baseline rate, which only affects the magnitude of the relative risk.) On the other hand, the probability that the average relative risk in the trials is less than unity is 0.99. The posterior mean of both θ and λ^{s+1} is -0.2. However, the posterior standard deviations are respectively 0.13 and 0.45.

This example illustrates how two-stage hierarchical models can be used to address study heterogeneity, one of the most relevant criticisms of meta-analysis, and to develop probability distributions for the quantities that are relevant in clinical decision making, that is, predictions for future patients. Ignoring study heterogeneity can be deleterious, and modeling heterogeneity is not difficult. Because conclusions can be sensitive to the prior distribution of τ, modeling heterogeneity requires careful consideration of what is a priori a reasonable amount of heterogeneity, based on past experience. For an example in breast cancer screening, see Berry (1998). We will return to this issue later in this chapter when we discuss sensitivity analysis.

4.5 Combining studies with continuous and dichotomous responses

A challenging problem in meta-analysis occurs when study responses, while similar, are not directly comparable. For example, some studies may report continuous outcomes, say the change in a functionality scale, while other studies consider a binary representation of a similar response, for example by recording whether there is an improvement in patient functionality. A simple approach in this situation is to dichotomize the continuous responses and proceed as in the simpler all-binary case, for example as in Section 4.4. This approach is practical, but it has limitations: for example, the choice of cutoff point for the dichotomization may be arbitrary, and a loss of information will occur in the dichotomization of the continuous response. The case study in this chapter will call for a strategy that overcomes both of these difficulties. Hasselblad and Hedges (1995) propose continuous scales based on the logits of the observed frequencies in the binary studies. These overcome the discretization problem in a nice way.

Here we discuss a Bayesian alternative, which has the added advantages of fitting conveniently within a hierarchical framework and leading to a simple solution to the problem of ranking treatments. In this section I present the Bayesian strategy in a simple case. The discussion is based on Dominici and Parmigiani (2000). The key step is to think of the binary responses as the result of dichotomizing some underlying unobserved continuous response. For example, whether there is an improvement in patient functionality can be approximately though of as a dichotomization of the change on a functionality scale, with yes corresponding to a positive change. The continuous scale can be used as a common underlying scale for all the studies. In this way we preserve the full information from the studies reporting continuous variables, we do not need to choose any arbitrary cutoff points, and we incorporate uncertainty arising from the heterogeneous nature of the responses. Technically, there is a problem in that this scale is missing in some studies and will have to be inferred from the observed binary response. But, as we have seen, MCMC methods are well suited to this task. Simulation-based methods for handling uncertainty about missing data in Bayesian analysis are discussed in detail by Tanner (1991) and Gelman et al. (1995).

4.5.1 A simulated example

By way of a simple concrete illustration of this idea, we now consider a simulated example, chosen to highlight the differences between the latent variable approach and alternative analyses that dichotomize the first two studies and then proceed as in the all-binary case. Consider four studies, indexed by s, each comparing a treatment arm ($t = 1$) with a placebo (or control) arm ($t = 0$). For each study, 20 observations were simulated in each arm. Studies 1 and 2 record a continuous response, while studies 3 and 4 record a binary response. Summaries of simulated data from the four studies are shown in Table 4.2.

We will use the notation y_{ti}^s for the observed response of the ith individual ($i = 1, \ldots, 20$) assigned to arm t in studies 1 and 2, and the notation x_{ti}^s for the binary outcomes corresponding to the observation in studies 3 and 4. To build a simple model for these data, assume that the continuous responses are normally distributed, that is

$$y_{ti}^s \mid \theta_t, \sigma^2 \sim N(\theta_t, \sigma^2), \quad t = 0, 1, \quad s = 1, 2,$$

where θ_0 and θ_1 are the placebo and treatment population means, and σ^2 is a common population variance. For studies 3 and 4, introduce the latent variables y_{ti}^s, where

$$y_{ti}^s \mid \theta_t, \sigma^2 \sim N(\theta_t, \sigma^2), \quad t = 0, 1, \quad s = 3, 4,$$

and $x_{ti}^s = 1$ whenever $y_{ti}^s > 0$. Therefore the binary observations have sampling distribution

$$P\{x_{ti}^s = 1 \mid \theta_t, \sigma^2\} = 1 - \Phi\left(-\frac{\theta_t}{\sigma}\right), \quad t = 0, 1, \quad s = 3, 4,$$

Table 4.2 Summary statistics from the simulated data with continuous and dichotomous responses. $\bar{y}_t^s$ and v_t^s are the sample mean and variance in arm t of study s.

Study	$\bar{y}_0^s$	v_0^s	$\bar{y}_1^s$	v_0^s
1	0.0218	0.0087	−0.0162	0.0075
2	0.0015	0.0103	0.1119	0.0105
	$\sum_i x_{0i}^s$	$\sum_i x_{1i}^s$		
3	12	8		
4	10	17		

where Φ is the standard normal cdf. It is convenient to indicate by $y_t^s = (y_{t1}^s, \ldots, y_{t20}^s)$ the vectors of continuous patient measurements in arm t of study s. In studies 3 and 4 this is not observed directly.

If the binary responses arise in practice as a dichotomization of the continuous responses, this model describes directly the data generation mechanism. Otherwise, the latent variables can be interpreted as hypothetical continuous outcomes consistent with the observed discretized outcomes. Using latent normal variables to model binary observation is discussed in greater generality by Carlin and Polson (1992) and Chib and Greenberg (1998). Because the point of this section is to illustrate the latent variables approach, we consider a fixed effects model, ignoring for the moment potential study heterogeneity.

When continuous and discrete responses are believed to differ systematically, an offset parameter α could be added to the model. The distribution of the continuous variables in studies 1 and 2 could be specified as

$$y_{ti}^s \mid \theta_t, \sigma^2 \sim N(\theta_t + \alpha, \sigma^2), \quad t = 0, 1, \quad s = 3, 4,$$

where α can be interpreted as the difference in measured efficacy between the two types of response. In order better to focus on the mechanics of the latent variable approach, we do not consider this case here.

In the model we have specified, all parameters are identified. In more general formulations, one may consider both treatment effects and variances to be study-specific. However, in studies with binary responses, we cannot make inference on both the mean and the variance of the underlying latent variables from the observed binary xs. Therefore, identifying restrictions are necessary, for example specifying a common variance σ^2 in all the studies.

Our Bayesian formulation is completed by specifying prior distributions for θ_0, θ_1, and σ^2. In the absence of more specific information, we choose conjugate, vague priors: $\theta_t \overset{\text{ind}}{\sim} N(0, 25)$, $t = 0, 1$, and $\sigma^2 \sim IG(a, b)$, independent of the θs. We use the parameterization of the inverse gamma with density function $f(x) = b^a x^{-a-1} e^{-b/x} / \Gamma(a)$ for $x > 0$, with mean $b/(a-1)$ and variance $b^2/(a-1)^2(a-2)$. We select hyperparameters $a = 3$, the integer value giving the most diffuse finite-variance prior distribution, and $b = 0.5$, selected to produce a mean of 1.

We can draw inferences on any of the unknowns using the joint posterior distribution

$$p(\theta_0, \theta_1, y_0^3, y_1^3, y_0^4, y_1^4 \mid \text{data}), \tag{4.1}$$

where data $= (y_0^1, y_1^1, y_0^2, y_1^2)$. Neither (4.1) nor its marginals are available in closed form. A practical choice for determining the marginal distribution, marginal probabilities, and other summaries of interest is again to draw a sample of values from (4.1). This can be done using a Gibbs sampler (Gelfand and Smith, 1990), based on partitioning the unknown parameters and missing data into groups and sampling each group in turn, given all the others. Observe that, if the y_{ti}^s are known for $s = 3, 4$, then posterior inference and simulation can be obtained easily using routine normal theory (Bernardo and Smith, 1994). The y_{ti}^s are, of course, unknown; however, given the data x_{ti}^s, the conditional distributions of y_{ti}^s given $x_{ti}^s = 1$ (0) are normal $N(\theta_t, \sigma^2)$ truncated to the positive (negative) values. With this in mind, we can write the full conditional distributions as follows. For the latent variables in studies 3 and 4:

$$y_{ti}^s \mid x_{ti}^s, \theta_t, \sigma^2, \text{data} \sim N(\theta_t, \sigma^2) I_{y_{ti}^s > 0}(y_{ti}^s) \qquad \text{for } i : x_{ti}^s = 1,$$
$$y_{ti}^s \mid x_{ti}^s, \theta_t, \sigma^2, \text{data} \sim N(\theta_t, \sigma^2) I_{y_{ti}^s \leq 0}(y_{ti}^s) \qquad \text{for } i : x_{ti}^s = 0.$$

For the parameters:

$$\theta_t \mid \sigma^2, \text{data}, \ y_0^3, y_1^3, y_0^4, y_1^4 \sim N(a_1, s_1^2)$$
$$\sigma^2 \mid \theta_0, \theta_1, \text{data}, \ y_0^3, y_1^3, y_0^4, y_1^4 \sim \text{IG}(a_1, b_1)$$

where $t = 0, 1$ and

$$a_1 = \left[\frac{1}{\sigma^2} \sum_{s=1}^{4} n_t^s + \frac{1}{s^2} \right]^{-1} \left[\frac{1}{\sigma^2} \sum_{s=1}^{4} n_t^s \bar{Y}_t^s + \frac{a}{s^2} \right],$$

$$s_1^2 = \left[\frac{1}{\sigma^2} \sum_{s=1}^{4} n_t^s + \frac{1}{s^2} \right]^{-1},$$

$$a_1 = a + \sum_{s=1}^{4} \sum_{t} n_t^s,$$

$$b_1 = b + \frac{1}{2} \sum_{s=1}^{4} \sum_{t} \sum_{i=1}^{n_t^s} (y_{ti}^s - \theta_t)^2,$$

where n_t^s is the sample size in arm t of study s.

Implementing the Gibbs sampler above, we can use sampled values from the chain to estimate the marginal densities of quantities of interests, such as the effect size difference $\theta_1 - \theta_0$. The posterior probability of a negative effect size difference, that is, $P(\theta_1 - \theta_0 \leq 0 \mid \text{data})$, is 0.021, while the 95% probability interval is $(0.004, 0.181)$.

It is interesting to contrast these results with those obtained by dichotomizing the continuous variables at 0 and then combining studies, which leads to a two-sample comparison of proportions. Our prior specification for the θs and σ^2 implies a U-shaped, approximately independent prior on such proportions. Replacing this prior with a product of noninformative priors of the form $1/p(1-p)$, the joint posterior on the proportions is approximated by a product of beta densities. The probability that the proportion of successes is greater in the treatment group than in the placebo is 0.076. The 95% equal-tail probability interval for the difference in proportions is $(-0.041, 0.263)$, which includes 0. While the two priors are not equivalent, they are sufficiently similar that we can attribute much of the difference in tail probabilities to the loss of efficiency resulting from the dichotomization of the observed continuous data. Similarly, the p-value of the two-sided Mantel–Haenszel test obtained by dichotomizing studies 1 and 2 with cutoff point at 0 is 0.1936. Therefore, at a level of 0.05, the latent variables approach leads to the conclusion that there is a significant difference between the treatment and the placebo effect, while the dichotomized approach, either in the Bayesian or frequentist formulation, does not. While this difference in testing outcomes naturally depends on the specific settings of the simulated example, the loss of efficiency it illustrates is general.

4.6   Migraine headache

4.6.1   Background and goals

Chronic migraine headache is a common condition, it is difficult to treat, and it has substantial impact on public health and productivity (Ziegler, 1990; Lipton and Stewart, 1993). A recent study of prevalence in the US population concluded that '8.7 million females and 2.6 million males suffer from migraine headache with moderate to severe disability. Of these, 3.4 million females and 1.1 million males experience one or more attacks per month' (Stewart et al., 1992). A wide range of drug and nondrug headache treatments are available, and there is wide disagreement about which are most effective (Pryse-Phyllips et al., 1997). The recent consensus conference of the Canadian Medical Association (Pryse-Phyllips et al., 1997) emphasized that migraine headache continues to be inadequately managed. Problems are particularly severe for prophylactic treatment, where a plethora of small clinical studies and trials offer conflicting evidence about treatment efficacy.

The Agency for Health Care Policy and Research (AHCPR) supported a large-scale systematic review of various categories of headache treatment. A team of clinicians reviewed the literature to select and abstract studies for a comprehensive evidence report. In our discussion we will focus on two types of drug therapy for migraine headache: (beta-blockers, and calcium-channel blockers. We draw data from the evidence reports by Goslin et al. (1998a; 1998b), which include 19 studies of four beta-blockers and eleven studies of four calcium-channel blockers. The drugs and their abbreviations are summarized in Table 4.3. Each study includes two or three treatment groups, so no study compares all the drugs directly. In 13 out of the 19 beta-blockers studies, and in 9 out of the 11 calcium-channel blockers studies, one of the experimental groups is given a placebo. The other studies consider drug-to-drug comparisons.

A meta-analysis of clinical trials can help bring existing evidence to bear on the controversy about which treatments are most effective. Often, because

Table 4.3 Summary of abbreviations for the migraine headache drugs examined in Section 4.6.

Calcium-channel blockers		Beta-blockers	
CB0	Placebo	BB0	Placebo
CB1	Verapamil	BB1	Propanolol
CB2	Nifedipine	BB2	Timolol
CB3	Flunarizine	BB3	Nadolol
CB4	Nimodipine	BB4	Metoprolol

of the relatively low response rate to prophylaxis, several treatments are prescribed in sequence to the same patient. To support better-informed choices of treatment sequences, it is important to provide a ranking of treatments, and uncertainty statements about this ranking. Also, despite the large number of published studies on the subject, relatively few medications for prophylactic treatment of migraine headache have been subjected to adequate clinical trial (Pryse-Phyllips *et al.*, 1997), and new trials will be necessary to resolve many of the remaining uncertainties. In our discussion we will present a model whose goal is to support clinical treatment decisions, and potentially help guide the planning of new trials. The critical statistical aspects of these goals are the synthesis of this complex and diverse information, the estimation of treatment effects on a common scale, and the relative ranking of treatments, both within classes and overall.

In developing a statistical model to address these goals we face several challenges. First, because individual studies only include a subset of treatments, and do not always include a placebo, we need to make indirect comparisons among treatments that may never have been tested together in the same trial.

Second, while most studies report continuous treatment effects for each treatment, some report only differences in effectiveness for pairs of treatments and others report only 2×2 contingency tables for dichotomized responses. Also, all the published responses that we have available have been previously standardized by dividing observed treatment differences by the estimated population subject-to-subject standard deviation within each study, to give them a common dimensionless scale (the primary sources reported effects on a variety of scales, including ordinal measures of well-being).

Finally, studies differ in the modalities of administration of the treatments, and in the criteria for defining migraine headache and admitting patients. On the other hand, the fact that the drugs' mechanisms are similar suggests that treatment effects are likely to be as well. Also, similarity is likely to be stronger within treatment classes than across classes. Because of the sparseness of the study/treatment matrix, and the relatively small size of some of the studies, it is important to exploit this similarity to 'borrow strength' from other studies in estimating each effect – borrowing more heavily from treatments of the same class.

Here we present an analysis that addresses these challenges using a combination of two techniques: hierarchical Bayesian modeling and latent variables modeling. Hierarchical modeling is used to model study heterogeneity and differential borrowing of strength within and across drug classes. In view of the previous paragraph), and the absence of study-specific covariate information to quantify effect variations, both studies and treatments present features that would make a traditional fixed effects model inadequate for the analysis. As we have seen in Section 4.4, hierarchical models offer a convenient way of modeling study heterogeneity. Latent variables modeling is used to account for study results having been standardized, some studies reporting only differences between treatments, and some studies reporting only dichotomized responses. We create a latent scale common to all studies for combining information from studies that report results in different forms. Our strategy will be similar in spirit to that of Section 4.5.

This model permits us to synthesize this heterogeneous information and to make inferences about treatment effects and the relative ranks of treatments, without ignoring important components of uncertainty. Estimation, ranking, model validation, and sensitivity analysis are all implemented through simulation-based methods. Our analysis is based on Dominici et al. (1999), who present a model sharing the same principles and motivation. Compared to what is presented here, their model is more complex in that it includes a third group of treatments, the biofeedback treatments. These are not drug treatments, and the borrowing of strength from the treatment effects requires an additional level in the hierarchical structure. Our discussion also differs in a number of small details of model, prior, and sensitivity analysis specification.

4.6.2 Data

The study effects and standard deviations from the 30 trials selected in the AHCPR preliminary evidence report (Goslin et al., 1998a; 1998b) are summarized in Table 4.4. Despite the relatively large number of trials, the data are sparse. Four of the treatments only appear in two trials.

The treatments under consideration are indexed by $t = 0, \ldots, 8$; $t = 0$ denotes the placebo groups in both drug classes. $T_{bb} = \{1, \ldots, 4\}$ is the set of indexes for the beta–blockers and $T_{cb} = \{5, \ldots, 9\}$ is the set of indexes for the calcium-channel blockers, both in the same order as in Table 4.4. The studies are indexed by $s = 1, \ldots, 30$, again in the order of Table 4.4. Each (t, s) combination is referred to as an arm. For each $s \in \mathcal{S}$, let $T^s \subset T$ be the set of treatments included in study s. For example, from rows 3 and 20 of Table 4.4, we see that $T^3 = \{0, 1, 2\}$ and $T^{20} = \{0, 7\}$.

Table 4.4   Reported study results for the 30 studies. Rows correspond to studies. Columns in each subtable correspond to treatments, with beta blockers in the top table and calcium-channel blockers in the bottom table. Entries in the tables are observed standardized treatment effects z. In parentheses are the standard deviations $\sqrt{q}$. A ○ indicates a crossover study. A ★ indicates studies reporting 2×2 tables. In study 5 there were 0 successes and 8 failures in the placebo group, and 6 successes and 14 failures in the treatment group. In Study 23 there were 14 successes and 15 failures in the placebo group, and 23 successes and 6 failures in the treatment group.

Study	n_s	BB0	BB1	BB2	BB3	BB4
1	52	−1.262 (1.3)	−0.687 (0.56)			
2	65	−0.016 (0.98)				0.508 (1.03)
3	240	−1.204 (1.21)	−0.874 (0.88)	−0.725 (0.87)		
4	58		−0.79 (0.92)		−1.34 (1.06)	
5	28	★			★	
6	94	−0.425 (1.03)		0.425 (0.97)		
7	32	○				0.497 (1) ○
8	40		−0.341 (1.02)			0.308 (0.98)
9	14	○		0.497 (1) ○		
10	64	−1.516 (1.1)	−0.953 (0.89)			
11	67		−1.611 (1.04)			−1.462 (0.95)
12	27		0.146 (1.03)		−0.158 (0.97)	
13	56		○			0.929 (1) ○
14	56	−2.033 (1)	−1.513 (1)			
15	55	−0.241 (1.28)	1.74 (0.71)			
16	34	−1.314 (1.18)	−0.902 (0.78)			
17	40		0.501 (1.11)			−0.751 (0.81)
18	30	○	0.497 (1) ○			
19	59	0.181 (0.93)				0.388 (1.07)

Study	n_s	CB0	CB1	CB2	CB3	CB4
20	42	0.23 (0.9)			0.449 (1.09)	
21	29	○			0.497 (1) ○	
22	20	−1.581 (1)		−1.423 (1)		
23	29	★			★	
24	24	−1.364 (1.12)	−0.987 (0.87)			
25	30	0.325 (1.17)			0.949 (0.8)	
26	28	0.069 (0.94)			0.526 (1.05)	
27	14	○	1.141 (1) ○			
28	78			0.408 (1.03)	0.556 (0.97)	
29	25				0.237 (1.25)	0.39 (0.63)
30	50	0.485 (1.2)				0.924 (0.82)

For studies with continuous outcome that include estimated individual treatment effects, we know the standardized treatment effect estimates z_t^s and the standardized variances q_t^s. These quantities are derived from original continuous data that are not available for our analysis, and will be denoted by Y. Specifically, we denote by Y_{ti}^s the response of the ith of the n_t^s individuals assigned to treatment t in study s, and by Y_t^s the sample mean for treatment t in study s. Then the sample variance for arm t of study s is

$$v_t^s = \frac{1}{n_t^s - 1} \sum_{i=1}^{n_t^s} (Y_{ti}^s - Y_t^s)^2,$$

and the pooled estimate of the variance in study s is a weighted average of the individual arms' sample variances, given by

$$v^s = \frac{\sum_{t \in \mathcal{T}^s} (n_t^s - 1) v_t^s}{\sum_{t \in \mathcal{T}^s} (n_t^s - 1)}.$$

The standardized quantities that are available to us are defined as

$$z_t^s = \frac{Y_t^s}{\sqrt{v^s}},$$

$$q_t^s = \frac{v_t^s}{v^s}.$$

This is a common strategy for reporting study effects in meta-analysis (Hedges and Olkin, 1985).

A small number of studies originally measured and reported an ordinal scale. These were subsequently transformed to a continuous scale using cumulative ranks. Only the transformed outcomes are available from the evidence report, so these studies are treated as continuous. The five studies indicated by o's in Table 4.4 adopted a crossover design, in which each patient receives both treatments at different points in time (Piantadosi, 1997). For those we only have available the estimated differences of standardized treatment effects, $z_t^s - z_{t'}^s$, between the two arms (usually between a treatment and a placebo), and the overall sample size $n_t^s + n_{t'}^s$. Study 27 (also indicated by a o in Table 4.4) reports only the differences between the two arms, but the design is a standard randomized trial. For brevity we will sometimes refer to these as *incomplete studies*. Finally, two studies (indicated by ⋆'s in Table 4.4) treat the response as dichotomous, and report 2×2 contingency tables. We use the notation x_t^s for the number of successes in arm t of study s.

4.6.3 Sampling distributions and latent variables

Our overall strategy consists of two main steps: creating a common scale for the response to treatment in all studies; and developing a hierarchical model for this common scale. For the continuous studies, and the so-called incomplete studies, the natural scale is given by the original, though unreported, values of Y. In crossover studies we introduce additional latent variables representing the unobserved effects in each of the arms. For the binary response studies we introduce, as in Section 4.5, latent normally distributed auxiliary variables and regard the reported binary responses as dichotomizations of the unobserved variables. In both cases our inference is based on an imputation of possible unreported responses for a hypothetical two-armed trial with continuous response, consistent with the abbreviated trial reports, and reflecting the uncertainty about the latent quantities.

The main structural assumption of our model is that the average response in arm (t, s) is the sum of two components: a study effect μ^s that represents the differences in responses across studies, and captures differing patient populations, protocol variations, and so on; and a treatment effect θ_t. This provides a reasonable flexible structure and at the same time allows comparisons of treatment effects across studies. In this setting the effect of a drug is assumed to be the same across studies. While it is technically feasible to model interactions between study and treatment efficacy, the data of Table 4.4 are too sparse to reliably estimate the resulting model. An alternative to interaction terms is hierarchical modeling of the treatment effects across studies, as was done in Section 4.4. Again, the sparsity of the data and the large number of drugs make it prohibitive to estimate such a model. For many drugs we would need to estimate a population of drug-specific effects based on two studies.

Our assumptions about the sampling distribution of the responses Y_t^s are as follows:

(i) the sample averages Y_t^s are approximately normally distributed with mean $\theta_t + \mu^s$ and variance σ_{ts}^2/n_t^s;

(ii) the sample variances v_t^s are approximately distributed as

$$\frac{\sigma_{ts}^2}{(n_t^s - 1)} \chi_{n_t^s - 1}^2.$$

The structure of the model for the complete continuous studies is shown in Figure 4.5. Each study provides information about as many θ_t as there are treatments in that study, and about the overall offset parameter μ^s which captures an overall shift in response for that study. In addition to the θ_t and μ^s, each study arm informs us about its own variance parameter. Because only standardized quantities are observed, there is no information about the original scale of the data. In other words, the empirical evidence remains equally likely if we multiply all the Ys by an arbitrary factor, say 2. Defining

$$\sigma_s^2 = \frac{\sum_{t \in T^s} (n_t^s - 1)\sigma_{ts}^2}{\sum_{t \in T^s} (n_t^s - 1)}$$

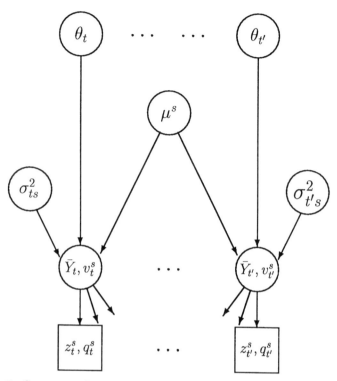

Figure 4.5 Summary of notation and conditional independence assumptions for the component model for complete study s. The squares with zs and qs are the observed standardized data; the circles with Ys and vs are the unobserved sufficient statistics for the latent normal response.

and $\gamma_t^s = \sigma_{ts}^2/\sigma_s^2$, the distribution of the observed zs and qs is independent of σ_s^2 given the γs. We can therefore set $\sigma_s^2 \equiv 1$ and interpret θ_t, μ^s, and σ_{ts}^2 as parameters of a model for the standardized sample averages Y_t^s/σ_s. By this convention the parameters σ_{ts}^2 are constrained to have a weighted average of unity within each study. In the incomplete studies there is no information about the γs, which will be assumed equal.

The hierarchical model of Figure 4.5 is specified in terms of unobserved variables Y_t^s, which are related to observed quantities in different ways for different studies. Anticipating our MCMC implementation, we give the conditional distribution of the latent Ys, given the reported observations. In the complete studies the unobserved v^s are distributed as

$$v^s \sim \frac{\sum_{t \in T^s} \sigma_{ts}^2 \chi_{n_t^s - 1}^2}{\sum_{t \in T^s} (n_t^s - 1)}.$$

Using v^s and the reported statistics, one can reconstruct all other unobserved statistics by way of $Y_t^s = z_t^s \sqrt{v^s}$ and $v_t^s = q_t^s v^s$.

The incomplete studies report only the standardized sample mean differences $(z_t^s - z_{t'}^s)$ and total sample sizes $(n_t^s + n_{t'}^s)$. Using the additional assumption that variances and sample sizes are equal in the two arms, v^s is distributed as

$$v^s \sim \frac{1}{(n_t^s + n_{t'}^s - 2)} \chi^2_{n_t^s + n_{t'}^s - 2}$$

and the reported difference is given by $\Delta_{tt'}^s \equiv (Y_t^s - Y_{t'}^s) = (z_t^s - z_{t'}^s)\sqrt{v^s}$. The conditional distribution of the missing sufficient statistics given the model parameters and the reported statistics is available from routine calculations with the normal distribution:

$$Y_t^s | \Delta_{tt'}^s \sim N\left(\mu^s + \frac{n_t^s \theta_t + n_{t'}^s \theta_{t'} + n_{t'}^s \Delta_{tt'}^s}{n_t^s + n_{t'}^s}, \frac{\sigma_s^2}{n_t^s + n_{t'}^s} \right).$$

For studies reporting only 2×2 tables we treat the data x_{ti}^s as indicators of the events $Y_{ti}^s > 0$, so that $P(x_{ti}^s = 1 | \theta_t, \mu^s, \sigma_{ts}) = \Phi((\theta_t + \mu^s)/\sigma_{ts})$. The constraint $\sigma_s^2 = 1$ still applies here. The conditional distribution of Y_{ti}^s given $x_{ti}^s = 1$ is $N(\theta_t + \mu^s, \sigma_{ts}^2)$, truncated to the positive values. When $x_{ti}^s = 0$, the same normal is truncated to the negative values.

4.6.4 Treatment and study variation

We now discuss the hierarchical model for the latent variables Y. The distributions of Y are controlled by treatment effects θ, the study effects μ and the arm-specific variances σ^2. We model hierarchically the θs and the μs, but not the σ^2s, as discussed. The overall structure of the model is shown in Figure 4.6.

The μs are modeled as independent and identically distributed random variables, independent of θ. To capture potential clustering in the distribution of the study effects we used a mixture of normal distributions. Each mixture component has an unrestricted variance, but the component means are constrained to give $E(\mu^s) = 0$. Based on diagnostic plots, discussed in Section 4.6.9, we choose a two-component mixture, that is

$$\mu^s | \omega, \alpha, \sigma_{\mu1}^2 \sigma_{\mu2}^2 \overset{\text{iid}}{\sim} \omega N(\alpha, \sigma_{\mu1}^2) + (1 - \omega)N\big(-\omega\alpha/(1 - \omega), \sigma_{\mu2}^2\big).$$

The resulting variance of the study effects is $\sigma_\mu^2 = \omega\sigma_{\mu1}^2 + (1 - \omega)\sigma_{\mu2}^2 + \omega\alpha^2/(1 - \omega)$. Here the mixture weight ω, the offset parameter α, and the component variances $\sigma_{\mu1}^2$ and $\sigma_{\mu2}^2$ are all unknown and must be estimated from the data. For identifiability, α is defined to be the positive one of the two component means, a constraint enforced through its prior distribution. This distributional specification, via the unknown parameters $\omega, \alpha, \sigma_{\mu1}^2$, and $\sigma_{\mu2}^2$, induces a dependence among the μ^s.

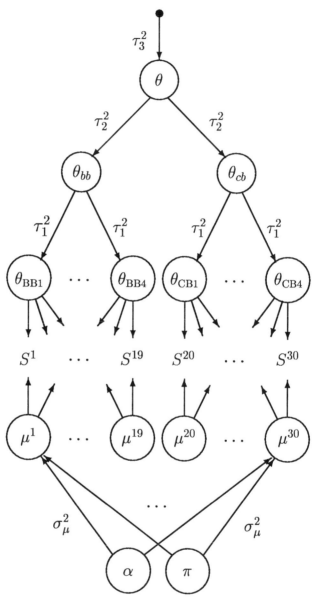

Figure 4.6 Summary of notation and conditional independence assumptions. $S^1, \ldots, S^{30}$ indicate the component models for each study, represented in Figure 4.5. Study and treatment effects are represented by nodes, while variance components are indicated next to the arc representing the corresponding conditional distribution. For example, τ_2^2 is the conditional variance of θ_{bb} given θ.

The response scale origins may differ across studies; to obtain a meaningful comparison, we set the placebo effect θ_0 to zero. For studies without a placebo arm, the study effect μ^s also embodies any difference in scale origins. The vector θ of treatment effects for nonplacebo arms is modeled as multivariate normal, with a dependence structure suggested by the clinical evidence of stronger similarity of treatments within classes, and weaker similarity for treatments across classes. We accomplish this with the two-level structure illustrated in the top portion of Figure 4.6.

Specifically, we define an overall mean parameter θ, and treatment class mean parameters θ_{bb} and θ_{cb}. Correspondingly, we have variance components τ_1^2 for the variation of effects within the same treatment class; τ_2^2 for the variation of class effects θ_{bb} and θ_{cb} with respect the overall mean parameter θ; and finally, τ_3^2 for the variation of the overall mean parameter θ with respect to its mean of 0. The marginal variance of each treatment effect is $\tau_1^2 + \tau_2^2 + \tau_3^2$. Our hierarchical specification implies correlations among the treatment effects. For treatments t, t' within the same class we have

$$\text{cor}(\theta_t, \theta_{t'}) = \rho_0 = \frac{\tau_2^2 + \tau_3^2}{\tau_1^2 + \tau_2^2 + \tau_3^2};$$

while for treatments in different classes we have

$$\text{cor}(\theta_t, \theta_{t'}) = \rho_1 = \frac{\tau_3^2}{\tau_1^2 + \tau_2^2 + \tau_3^2}$$

As desired, $0 \leq \rho_1 \leq \rho_0 \leq 1$.

Introducing these parameters extends the more common model where all treatment effects are drawn from the same distribution, which corresponds to setting $\rho_0 = 0$, or $\tau_2^2 = \tau_3^2 = 0$. Another special case is the separate analysis of the two treatment classes, with common variance τ_1^2 in all classes, which corresponds to $\rho_0 = 1$ and $\rho_1 = 0$, or $\tau_2^2 \to \infty$ and $\tau_3^2 = 0$. Our correlation structure generates differential borrowing of strength in the effect size estimates within and across classes. Because one of the main goals is to make inferences about relative ranking, it seems appropriate to incorporate clinical knowledge about the ensemble of treatments. In a different context, such as deciding whether to approve an individual treatment, other approaches may be more appropriate.

In summary, the distributional assumptions of our latent variables and hierarchical model parameters are as follows:

$$
\begin{aligned}
\theta_{bb}, \theta_{cb} &\overset{iid}{\sim} N(\theta, \tau_2^2) \\
\theta_t &\overset{ind}{\sim} \begin{cases} N(\theta_{bb}, \tau_1^2), \, t \in \mathcal{T}_{bb} \\ N(\theta_{cb}, \tau_1^2), \, t \in \mathcal{T}_{cb} \end{cases} \\
\mu^s &\overset{iid}{\sim} \omega N(\alpha, \sigma_{\mu 1}^2) + (1 - \omega) N\left(-\omega \alpha/(1 - \omega), \sigma_{\mu 2}^2 \right) \\
Y_t^s &\overset{ind}{\sim} N(\theta_t + \mu^s, \sigma_{ts}^2/n_t^s) \\
v_t^s &\overset{ind}{\sim} \left(\sigma_{ts}^2/(n_t^s - 1) \right) \chi_{n_t^s - 1}^2.
\end{aligned}
\tag{4.2}
$$

Here each distribution is conditional on all the parameters higher in the list and hierarchy, although it may not depend on all of them as a result of the conditional independence assumptions represented in Figure 4.6.

While both this analysis and that of tamoxifen studies in Section 4.4 use hierarchical models, they differ in two important respects which highlight the flexibility of hierarchical models in this context. The first is that in the headache application the study effects are dependent, while in the tamoxifen application each study had a separate baseline success rate in the control arm and there was no borrowing of strength across studies for those parameters. The reasons for this difference are the much larger number of studies in the headache application and the smaller sample size of most of them. Both strategies could be defended in both analyses. The second is that the treatment effects are assumed to be constant across studies in the headache example, while they are allowed to be heterogeneous and modeled hierarchically in the tamoxifen application. In the headache example the hierarchical model describes variation of the treatment effect of similar drugs within the same class, while in the tamoxifen example the hierarchical model describes variation of the treatment effect of the same drug in different studies.

4.6.5 Prior distributions

For a Bayesian analysis of this model we must specify distributions for all unknown parameters not included in the list (4.2). We will refer to these as priors, even though it is somewhat arbitrary to define the boundary between prior and likelihood in multilevel hierarchical models. For example, it is not unusual to think of the distributions on θ_{bb} and θ_{cb} as part of the prior as well.

An attractive and practical approach for choosing prior distributions on high-level parameters in complex hierarchical models such as this is to specify a dispersed but proper baseline prior distribution, and to supplement the baseline analysis with additional sensitivity analyses. In Section 4.6.10 we present sensitivity analyses for four possible departures from the following baseline prior specification:

Overall effect mean	$\theta \sim N(0, \tau_3^2)$	
Variance (treatment effects)	$\tau_j^2 \sim \mathrm{IG}(a_j, b_j),$	$j = 1, 2$
Variance (study effects)	$\sigma_{\mu i}^2 \sim \mathrm{IG}(a_\mu, b_\mu),$	$i = 1, 2$
Offset (study effects)	$\alpha \sim N(0, b_\alpha) I_{\alpha > 0}$	
Mixture weight (study effects)	$\omega \sim U(0, 1)$	
Variance (individual effects)	$\sigma_{ts}^2 \sim \mathrm{IG}(a_\sigma, b_\sigma),$	$t = 1, \ldots, 8$
		$s = 1, \ldots, 30.$

The as and bs are fixed hyperparameters.

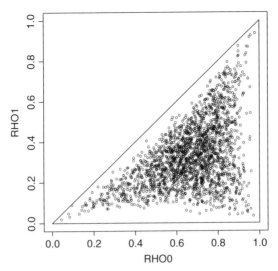

Figure 4.7 A sample from the joint prior probability distribution of the correlations ρ_0 and ρ_1 in the baseline specification.

Unless otherwise implied by the hierarchical structure, we used independent prior distributions. We use the parameterization of the inverse gamma with density function $b^a x^{-a-1} e^{-b/x} / \Gamma(a)$ for $x > 0$, with mean $b/(a-1)$ (for $a > 1$) and variance $b^2/(a-1)^2(a-2)$ (for $a > 2$). For our baseline analysis we selected hyperparameters $a_j = a_\mu = 3$, the integer value giving the most diffuse finite-variance prior distributions to our variance components, and $b_j = b_\mu = 2/3$, to ensure that the marginal expected treatment effect variance is $\tau^2 = \sum_{j=1}^{3} \tau_j^2 = 1$, appropriate for these standardized quantities, apportioned equally to the three levels of the hierarchy. A sample from the resulting distribution of the correlations ρ_0 and ρ_1 is shown in Figure 4.7. The prior distributes mass throughout the region. We assign the study effect offset α a half-normal distribution with variance $b_\alpha = 3$, and give improper reference distributions $(1/\sigma_{ts}^2)$ to the study-specific individual effect variances, since the sample sizes are sufficient to support a noninformative analysis of these parameters. We adopt the conservative approach of using prior mean 0 for the overall treatment and placebo effect mean θ, suggesting no prior information that any of the treatments is better or worse than the placebo.

4.6.6 Computing

We evaluate the posterior distribution of the parameters and unreported observations is Markov chain Monte Carlo simulation (Tanner and Wong, 1987; Tierney, 1994). Most of the full conditional distributions, including those of the latent sufficient statistics, are available in closed form and simple to sample from, making the Gibbs sampling variation of MCMC (Gelfand and Smith, 1990) viable. The identifying constraint $\sigma_s^2 = 1$ makes closed-form expressions for the study-specific variances unavailable, so we use a Metropolis step (Tierney, 1994). The normal mixture parameters for the study effects are updated by augmenting the data with a vector of latent mixture component indicators, as done by Diebolt and Roberts (1994) and West and Turner (1994) for their unconstrained cases. The constraint $E(\mu^s) = 0$ is linear in α and so the full conditional distribution of α is normal, truncated to the positive half-line. At each iteration the latent variables are simulated from the distributions given in Section 4.6.3.

It is important to check graphical diagnostics for convergence of the chain, as implemented, for example, in CODA (Best et al., 1995). In this application, the MCMC sampler mixes well, and a few hundred iterations typically suffice for convergence. Because iterations are not expensive, we present results based on a subset of 6000 equally spaced draws from the last 30 000 iterations of a single chain of 35 000 iterations.

4.6.7 Results

Our objective is to make inferences about treatment effectiveness, identifying which treatment class seems best and how well each treatment works, by computing the posterior distribution of the treatment effect vector θ, given the reported values z_t^s and q_t^s. The posterior distributions of the treatment effects θ's are shown in Figure 4.8. All θs have positive posterior means, indicating effective treatments. Overall, the probability that none of the treatments has a positive effect is smaller than 0.0002; the probability that at least one of the treatments has an effect larger than 1 is 0.46; the probability that all treatments have a positive effect is 0.35.

To emphasize the contribution of individual studies and to validate the assumption of homogeneity of effects across studies, Figure 4.8 also shows study-specific summaries $Y_t^s - \mu^s$. These are the quantities that contribute to the estimation of the treatment effect within each study. Because the μ^s are unknown, we replaced them with their posterior means. So each corresponds to a study including the treatment in question, and is positioned at the posterior mean of $Y_t^s - \mu^s$.

An important contribution of a meta-analysis to patient management is to quantify how much is known about which treatments are best. This can be expressed by a probability distribution over possible rankings. In addition to supporting treatment decisions, an accurate assessment of uncertainty may help guide the planning of new trials. In a simulation-based approach the ranking problem can be solved easily even in the presence of complex random effects and missing data that may induce significant dependencies in the joint distribution of the effects under comparison. Ranking is done simply by reporting the empirical frequency with which each treatment is

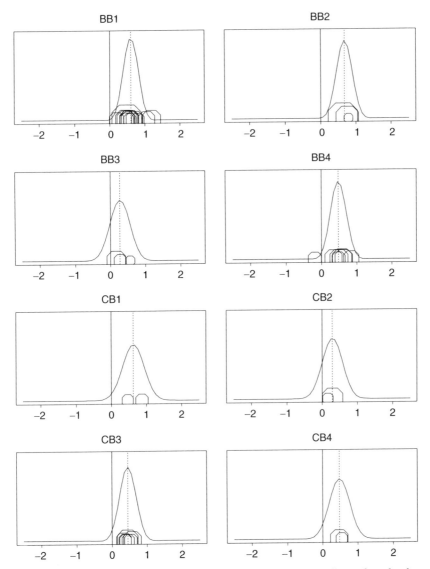

Figure 4.8 Marginal posterior distribution of treatment effects for the beta-blockers (top four) and calcium-channel blockers (bottom four) classes. Labels identify treatments; each circle corresponds to a study that includes the treatment in question, and is positioned at the posterior mean of $Y_t^s - \mu^s$; areas of circles are proportional to the sample sizes n_t^s. Posterior means and no-effect points are indicated by vertical dashed and solid lines, respectively.

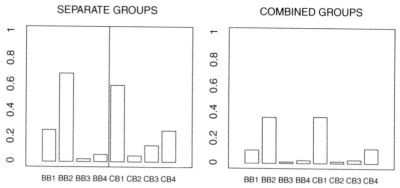

Figure 4.9 Probabilities that each treatment effect is the largest, by treatment class (left), and all treatments combined (right). Substantial uncertainty remains about which treatments are most effective, both within classes and overall.

better than its competitors within the same treatment group. In Figure 4.9, we show the probability that each treatment effect is the largest. To convey a sense of the remaining uncertainty about the choice of the best treatment, we also report an overall ranking. Labeling the drugs in Figure 4.9 from left to right as 1 to 4, the most probable ranking is $1, 4, 3, 2$, with an estimated posterior probability of 20.6%, followed by $4, 1, 3, 2$ and $4, 3, 2, 1$, with estimated posterior probabilities of 15% and 11%, respectively. None of the studies considered here compares treatments across classes, so that our comparisons across classes rely heavily on assumptions about how populations, and particularly placebo groups, compare across studies.

A comparison of the overall class effects θ_{bb} and θ_{cb} helps us address the issue of which drug class is likely to perform better. Figure 4.10 displays histograms of samples from the posterior distributions of the overall class effects, along with the corresponding prior densities. There is substantial overlap between the two posterior distributions, as expected, suggesting that considerable uncertainty remains about which class of treatments is most effective; in each case the prior is considerably more diffuse than the posterior, suggesting that the data offer evidence about treatment means.

In contrast, the data provide relatively little information about the correlations. The posterior distribution, shown in Figure 4.11, concentrates on high values for the within-class correlation coefficient ($E[\rho_0 \mid \text{data}] = 0.774$, with posterior interquartile range (IQR) $(0.674, 0.842)$), and moderate values for the related-class coefficient ($E[\rho_1 \mid \text{data}] = 0.394$, with posterior IQR $(0.274, 0.505)$); recall that the prior distributions had means $E[\rho_0] = 0.666$, $E[\rho_1] = 0.333$, with prior IQRs $(0.569, 0.788)$ and $(0.207, 0.419)$, respectively. In the figure, dashed vertical lines represent prior means, and solid vertical lines represent posterior means.

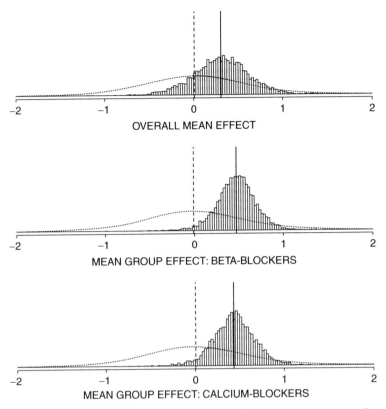

Figure 4.10 Posterior distribution of the overall effect θ and class effects θ_{bb} and θ_{cb} for the two treatments classes. The dotted lines are the prior distributions. Posterior means and no-effect points are indicated by vertical solid and dashed lines, respectively.

The approach and results presented here can potentially provide valuable input for the design of new trials. The treatment rankings can assist in choosing treatment arms; the treatment effect estimates can help in selecting sample sizes. Our hierarchical model can also provide the basis for more formal simulation-based optimal design. For any fixed choice of treatment, one can simulate from the predictive distribution of $Y_t^{s'}$ in a future study $s' \notin \mathcal{S}$. Simulated values can in turn be used to evaluate the information provided by a trial design (Müller, 1998). Incorporating two treatment classes within the same model enables one to consider future trials comparing treatments across classes.

4.6.8 Comparisons with alternative approaches

We now consider a comparison of our results with those obtained using two alternative model specifications. The first is a separate hierarchical analysis of

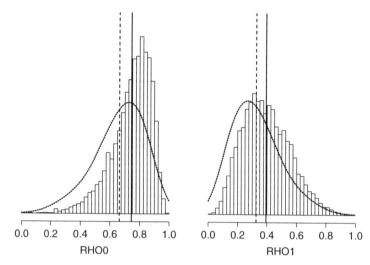

Figure 4.11   Prior and posterior distributions of correlation coefficients ρ_0 and ρ_1 of treatment effects within and across drug classes.

the two treatment classes, without the constraint of a common τ_1^2. In this case there is generally less shrinkage toward a common mean for the biofeedback treatments, and slightly higher posterior standard deviations. The variance component τ_1^2, measuring the heterogeneity of treatment effects within classes, is estimated with substantially greater precision in the combined analysis, where the parameter is common across classes. Because of the small number of treatments, it is important to incorporate information about the variability of effects in similar classes of treatments when estimating variances of treatment effects. Ranking within classes is not changed substantially. Ranking across classes is not possible in this analysis.

The second comparison is with a linear mixed model with study-specific random effects, estimated using the maximum likelihood approach proposed by Hasselblad (1998). Dichotomous outcomes measures were converted to effect sizes using the method of Hasselblad and Hedges (1995). This results in an estimate of the difference in effect sizes, so these studies are represented as contrasts, rather than effects, in modeling the arm-specific means. Crossover studies are treated similarly. Figure 4.12 compares the point estimates obtained using the two approaches. The point estimates are close, differing by less than 0.3. The Bayesian model systematically shrinks the estimates more strongly toward the center of the distribution of estimates. Shrinkage occurs in differing degrees to different effect sizes, the effect sizes that are less accurately estimated being pulled further in. This determines a reversal of ranking between verapamil and timolol for the most effective treatment and between nadolol and nifedipine for the least effective. The Bayesian model also features smaller variances reflecting the correlation within the

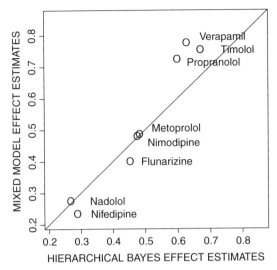

Figure 4.12 Comparison of Bayesian hierarchical effect size estimates (horizontal scale) and maximum likelihood effect size estimates using a linear mixed model.

treatment classes. However, the Bayesian model reflects the uncertainty from the incomplete studies, which tends to increase the variance. The two effects are roughly balanced for treatments in incomplete studies, but in other treatments the effect standard deviations can be reduced by a factor of 2. The linear mixed model approach does not lend itself easily to generating ranking probabilities.

Our results are consistent with those of a non-Bayesian meta-analysis with study-specific random effects, when combining the evidence from relatively large studies. However, they differ in the smaller studies by shrinking the study-specific reported treatment effects towards the mean effect for similar studies and towards the skeptical prior mean effect of zero. As a result of shrinkage, our analysis leads to conclusions that are both more precise in estimating individual treatments and more conservative in drawing conclusions about ranking. We regard this as a scientific justification for the medical community's natural reluctance to be swayed by striking results from small studies, and as a systematic tool for basing inference on a growing body of evidence.

4.6.9 Graphical model checking

When building a complex multilevel model, it becomes especially important to check the validity of assumptions about latent variables and parameters at higher levels in the hierarchy. An effective strategy is to inspect sequences of diagnostic plots of summary measures from the full conditional distributions.

Ideally, summaries should constructed to be conditionally independent and identically distributed if the model assumptions are satisfied. Here we discuss and illustrate this strategy in the context of validating model assumptions about the study effect distributions, and about the overall fit of the model. Another illustration of this idea is presented by Meulders *et al.* (1998).

We begin by considering the distribution of the study-specific random effects μ^s shown in Figure 4.13. The offset parameter α has posterior mean 0.27 and IQR $(0.13, 0.39)$; the mixture weight ω has posterior mean 0.44 and IQR $(0.36, 0.52)$. The two component-specific variances $\sigma^2_{\mu 1}$ and $\sigma^2_{\mu 2}$ have posterior means 1.00 and 1.01, and IQR's $(0.88, 1.11)$ and $(0.65, 1.25)$, respectively. Figure 4.13 illustrates the distributions of the individual study effects μ^s within the three treatment classes. Effects can be large, and are clearly heterogeneous across studies. Studies providing only treatment differences offer no direct evidence about individual study effects, leading to the most widely dispersed distributions. Studies reporting 2×2 tables also have dispersed distributions of random effects, but these can stray far from 0 for informative studies; for example, study 5 has a smaller study effect than study 23, as a result of its lower success rate in the placebo group. Study 15 has the largest positive study effect. Study 11 has one of the lowest study effects. This reflects the fact that both the BBI and the BB4 arms of that study have a low response, while they generally perform well elsewhere (Figure 4.8).

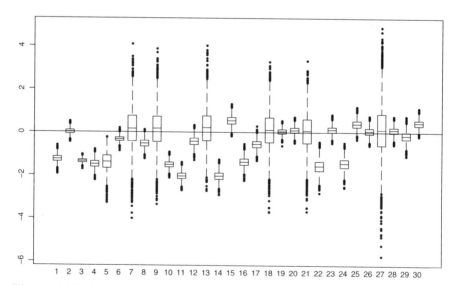

Figure 4.13   Posterior distribution of study effects. Study numbers are listed under each boxplot. Studies 7, 9, 13, 18, 23, and 27 report the estimated differences in effect sizes between the two arms. Studies 5 and 23 report 2×2 contingency tables.

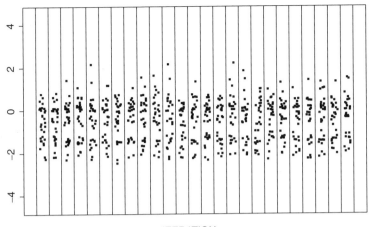

ITERATION

Figure 4.14 A random sample of distributions of the diagnostic summaries m. Each vertical stripe corresponds to one iteration in the Markov chain. At most iterations, the ms exhibit clustering into two distinct subgroups, supporting the choice of a two-component mixture for the distribution of the study effects.

When updating the study effects μ_s, given all other unknowns in the model, the data-dependent quantities

$$m_s = \frac{\sum_{t \in \mathcal{T}^s} (Y_t^s - \theta_t) n_t^s / \sigma_{ts}^2}{\sum_{t \in \mathcal{T}^s} n_t^s / \sigma_{ts}^2}$$

play the role of the observed sample means. For each iteration of the chain we can use the empirical distribution of the m's to assess the validity of the posited distribution of the μ's, much as when validating a likelihood function based on an empirical sample. We generated repeated samples of distributions of m's (Figure 4.14). Recurring features are skewness to the left and heavy tails. Inspecting these distributions and Kolmogorov–Smirnov statistics quantifying their nonnormality led us to replace the simple normal model specified initially with the two-component mixture adopted here. Direct inspection of the distributions of μs would not be as sensitive a diagnostic for two reasons: draws would be from the marginal posterior distribution of the μs rather that the conditional we are validating; and draws would depend directly on the current distributional assumption about the μs. We examined similar plots for the treatment effects. Because of the relatively small number of treatments, the data do not provide detailed information about the shape of the higher-level distribution for which the normal assumption appears to fit well.

To assess the overall fit of the model we considered the latent residuals $\epsilon_t^s = Y_t^s - \theta_t - \mu^s$ and studied their posterior predictive distributions. At each iteration of the chain we simulate data Y_t^s conditionally on the current

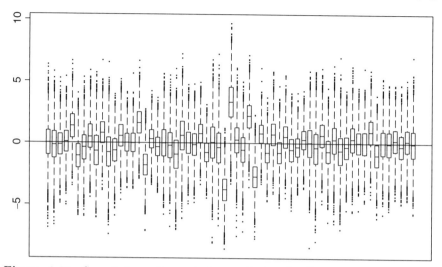

Figure 4.15   Sample from the posterior predictive distributions of the residuals. Each boxplot corresponds to an arm. Studies are ordered from left to right. Within each study, arms are in the same order as in Table 4.4. The vertical scale is $\eta_t^s = (Y_t^s - \theta_t - \mu^s)\sqrt{n_t^s}/\sigma_{ts}$, the standardized deviation of the latent sufficient statistic from its predicted mean. When 0 is in the tail of one of these distributions, the model does not fit well in that study arm.

values of the model parameters, and compute the resulting residuals; at the end of the MCMC simulation the collection of simulated residuals is a sample from their unconditional predictive distribution. For 0 to lie in the tail of one of these distributions is an indication that the model does not fit well in that study arm. We use boxplots to assess this informally, but more formal methods, such as predictive p-values, are also available (Meng, 1994; Rubin, 1984). To accommodate the different variances and sample sizes in the studies, we study the standardized latent residuals $\eta_t^s = \epsilon_t^s\sqrt{n_t^s}/\sigma_t^s$.

Figure 4.15 shows a sample of η_t^s from all arms. The fit of the model is very close to the data, with all boxplots including 0 within their whiskers. The predictive distributions display the heavy tails and other departures from normality that are characteristic of mixing over complex model specifications. In all cases 0 is within the whiskers of the boxplot, indicating that 0 is not an outlying observation in the distribution of residuals. There are three consecutive pairs of residuals in which the middle 50% of the boxplots are fully on opposite sides of 0. These correspond to studies 14, 31, and 33. In study 33, for example, the two arms have effects that are both discordant with the general tendency of the same treatments in other studies.

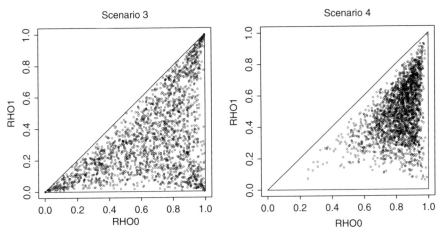

Figure 4.16 Samples from the joint prior probability distribution on the correlations ρ_0 and ρ_1 in scenarios 3 and 4.

4.6.10 Sensitivity to prior specification

The influence of our choice of prior hyperparameters can be addressed using a sensitivity analysis. Our strategy is to select four alternative scenarios for departure from our baseline hyperparameter prior distributions of Section 4.6.5, and evaluate under each of them the outcome of primary interest, in this case the ranking of treatments. The four scenarios are:

1. As the baseline, except that the study effects variance is set to be larger, by specifying $a_\mu = 6/5$ and $b_\mu = 20$ for both mixture components;
2. As the baseline, except that the study-specific variances are set to be close to unity, by specifying $(a_\sigma, b_\sigma) = (11, 10)$;
3. As the baseline, except the correlations ρ are given highly dispersed U-shaped prior distributions, by specifying $a_1 = a_2 = a_3 = 6/5$ and $b_1 = b_2 = b_3 = 1/15$;
4. As the baseline, except the variance component means increase with their place in the hierarchy. The specification is $a_1 = a_2 = a_3 = 3$ and $(b_1, b_2, b_3) = (1/3, 2/3, 1)$. This choice preserves the property that $E[\tau^2] = 1$, which is demanded by the standardization of the effect sizes.

Samples from the joint prior distribution on the correlations ρ_0 and ρ_1 in scenarios 3 and 4 are shown in Figure 4.16. Scenario 3 is nearly uniformly distributed, while scenario 4 favors higher correlations, and therefore a greater borrowing of strength from treatment to treatment.

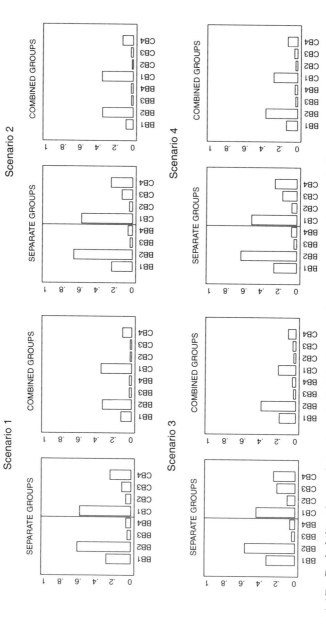

Figure 4.17 Probability that each treatment effect is the largest, by treatment class, and all treatments combined, under the four scenarios considered in our sensitivity analysis.

The probabilities that each treatment effect is the largest, by treatment class, and all treatments combined, under the four scenarios considered in our sensitivity analysis, are shown in Figure 4.17. The ranking of treatments is stable across these four scenarios, and therefore it does not appear that the conclusions about ranking are sensitive to the prior specification.

5

Decision trees

5.1 Summary

In this chapter we illustrate the modeling of multistage decision problems using decision trees. In a multistage problem, several decisions are made at different points in time, on the basis of different information. One critical aspect of this is that, as in a chess game, decisions made at one point may affect both the worthiness of options at future points, and the information available when making future decisions. A typical instance is the decision whether to perform a diagnostic test or a staging procedure to aid future treatment decisions.

We exemplify this in our case study of axillary lymph node dissection, a surgical procedure commonly used in early breast cancer. After giving the clinical background, we consider a simplified scenario in which the surgery is used solely for staging purposes, that is, for learning about the severity of the patient's cancer. This scenario will serve as a tutorial on decision trees and as prototypical example of a general paradigm for assessing the value of diagnostic information (DeGroot, 1984; Sox et al., 1988). We then give a more complete analysis of axillary lymph node dissection. Our discussion builds on a model developed in Parmigiani et al. (1999). While the structure of the model is similar, most of the relevant parameters have been updated. Also, the present discussion incorporates parameter uncertainty in a systematic way. An alternative decision analysis of the same surgical procedure is discussed by Velanovich (1998).

5.2 Axillary lymph node dissection in early breast cancer

Treatment of early-stage breast cancer starts with an initial surgical procedure, to remove the tumor and some of the surrounding tissue. This is often followed by additional (adjuvant) systemic treatments, such as chemotherapy and hormonal therapy, to control breast cancer cells that may have spread elsewhere in the body (Harris, 1996). Surgical options include removal of the breast, via one of the variants of mastectomy, and breast conservation therapy, via a lumpectomy. Lumpectomy is a partial form of mastectomy and conserves varying degrees of breast tissue. Axillary lymph node dissection (ALND) is also a surgical procedure consisting of the removal of the axillary lymph nodes. It is performed routinely as part of a mastectomy. In breast conservation patients ALND is performed often, but technically it is a distinct procedure, often requiring a separate incision.

ALND has been an integral part of the management of early breast cancer since the work of Halsted (1894–1895). Initially, the purpose of ALND was to remove the likely first area of potential tumor spread from the primary lesion in the breast. It was thought that this surgical approach would result in a higher cure rate than if only the breast were removed. Attempts at less extensive surgery began in the 1930s and 1940s, primarily in Europe. Clinical trials conducted in the 1970s and 1980s demonstrated that less extensive surgery can lead to comparable survival rates. At a turning point, National Surgical Adjuvant Breast and Bowel Project (NSABP) study B-04 (Fisher et al., 1985), did not show any improvement in survival as a result of ALND, despite a relatively high frequency of clinically node-negative patients with axillary lymph node metastases. Today, breast conservation therapy has become increasingly common, its equivalence to radical mastectomy or modified radical mastectomy having been demonstrated in randomized trials (Morris et al., 1997). As a result, ALND for many patients is now a separate operation rather than part of a mastectomy.

In parallel, effective radiation therapies and adjuvant systemic therapies have been developed. Total mastectomy and irradiation of peripheral lymph nodes was eventually shown to be comparable in outcome to radical mastectomy, leading to the conclusion that radiation therapy could successfully control subclinical disease in lymph nodes (Keynes, 1936; Johansen et al., 1990). The concept of breast cancer as a systemic disease has led to the increased use of adjuvant therapy in a wide range of patients, benefits having been demonstrated in node-negative as well as node-positive patients (Early Breast Cancer Trialists Collaborative Group (EBCTCG), 1992; 1998a).

It is now generally thought that positive axillary nodes are primarily a predictor of occult systemic spread, but not necessarily the source of systemic metastases. ALND is therefore viewed primarily as a staging procedure which conveys prognostic information and may guide decisions regarding adjuvant systemic therapy. For example, when adjuvant systemic therapy was introduced, a renewed emphasis was placed on ALND as a means to select patients with a worse prognosis for more aggressive adjuvant treatment. Today, however, the relative benefits of adjuvant therapy appear to be similar in node-negative and node-positive women. Adjuvant systemic therapy is suggested as appropriate for all node-negative patients with a primary tumor larger than 1 cm (St Gallen recommendation (Davidson and Abeloff, 1994)), calling the role of the ALND into question.

Finally, the distribution of the extent of disease in newly diagnosed patients has changed, as the use of mammography and other early detection techniques has grown and public awareness of breast cancer has increased. Many more patients are encountered with small primary tumors. These patients have a low probability of nodal involvement (Carter *et al.*, 1989), and a good prognosis even if a small number of positive nodes are found (Quiet *et al.*, 1996). Thus the survival benefits from ALND are likely to be small in this group. Also, elderly patients, particularly those with hormone receptor positive tumors, will often be treated with tamoxifen irrespective of nodal status, because of the reluctance to use adjuvant chemotherapy in this age group.

5.2.1 *Why a decision model?*

This complex scenario has given rise to wide disagreement on the routine use of ALND (Cady, 1996; Moore and Kinne, 1996; Silverstein *et al.*, 1994; Epstein, 1995; Recht and Houlihan, 1995). To summarize and somewhat simplify, the reasons why ALND is currently performed in breast conservation patients include the following:

1. To guide future action. Specifically, ALND is useful when it results in an alteration of a patient's plan of adjuvant therapy in the absence of ALND and that different plan improves survival or decreases morbidity. While the relative benefits of adjuvant therapy do not appear to depend on nodal status, the prognostic information conveyed by the results of ALND may still be critical in decision making, as a patient with a worse prognosis may be more inclined to undergo a more aggressive form of treatment.

2. To provide local control of the disease. There is evidence that ALND patients survive longer than patients receiving no axillary treatment. Axillary irradiation (XRT) appears, however, to provide local control and survival benefits (if any) comparable to ALND in patients with clinically uninvolved lymph nodes. A recent comprehensive meta-analysis of clinical trials by EBCTCG (1995) indicates a mortality reduction of about 4% for ALND over radiation therapy, with about 4% standard deviation. While not statistically significant, a 4% mortality reduction, if true, would be of substantial clinical significance, especially for younger women.

3. To increase the accuracy of the prognosis. To many patients the prognostic information gained from ALND may be valuable separate from the impact it has on local control and on subsequent therapeutic decisions. The life expectancy of a woman with a large involvement of lymph nodes can be half or less compared to that of a woman with no nodes involved. The value of this information is linked to personal decisions and sometimes motivated solely by the 'desire to know'. This is difficult to quantify reliably, and we will not incorporate it directly in our model.

These benefits must be considered in light of the negatives, which include:

1. Morbidity. Patients report several side effects, including decreased range of motion of the shoulder (10–20%), numbness in the distribution of the intercostal brachial nerves (80%), arm edema (2–30%), and breast edema (15–50%) (Warmuth et al., 1998).
2. Financial costs of the procedure (Orr et al., 1999).

In this chapter we use the methods of decision analysis to investigate these trade-offs quantitatively. Not all dimensions of this problem can be reliably quantified, and we will not attempt to quantify everything, but we hope to illustrate how quantification can provide concrete help to decision making.

Our focus is on patients who are candidates for breast conservation therapy, since ALND is a separate operation for these patients, while it is not so for patients undergoing mastectomy, for whom it is typically an integral part of the procedure. In addition, we restrict attention to patients whose clinical examination of the axilla is negative, thus excluding patients who have palpable nodal involvement. There is wider agreement that ALND is appropriate for these patients. The decision to perform a node dissection and the selection of appropriate adjuvant therapy are made in the light of patient characteristics, namely age, estrogen receptor (ER) status (either positive or negative), and size of the primary tumor, as benefits are likely to be different for these groups (Lin et al., 1993).

In Section 5.3 we illustrate the basics of decisions trees, using the simplest decision tree that is sufficient to capture the essential traits of a medical diagnostic problem. While we use ALND as the main example, we simplify the situation by assuming that ALND is only performed to aid subsequent choice of adjuvant therapy. The resulting example is the archetype of decision trees for evaluating the clinical utility of a diagnostic test, and will give us the opportunity to discuss a simple approach to quantifying the practical value of the information provided by a medical diagnostic procedure. In Section 5.4 we consider a more realistic treatment of the ALND decision model, incorporating direct effects of ALND on both local control and quality of life, and introducing a range of patient types. We discuss in detail how we obtained the estimates that are then used to solve the decision tree.

In our simplified example, we consider a cohort of 60-year-old patients, ER positive, with a primary tumor of size 0.75 cm. In the general case, we look again at the same cohort in detail, and then examine all covariate combinations.

5.3 A simple decision tree

5.3.1 Nodes, branches, and leaves

Because knowing the result of ALND might later lead to a more appropriate choice of adjuvant therapy, we need to consider two related decisions: the choice of either ALND or XRT; and the choice of adjuvant systemic therapy, possibly based on the results of the ALND. Decision problems with multiple related choices can often be addressed using decision trees. A decision tree is a graphical technique that allows us to break a complex decision problem down into simpler interconnected decision problems. It provides ways of representing and communicating decision models, as well as ways of finding and representing recommendations that may depend on past medical history.

Like a real tree, a decision tree has nodes and branches. Nodes can be of two kinds: *decision nodes*, representing choices the decision maker (DM) will have to make; and *chance nodes*, representing events that will occur in the future and may affect the DM's subsequent choices. Branches stemming from decision nodes represent alternative options, while branches stemming from chance nodes represent alternative outcomes.

In a decision tree, decision nodes are represented by squares or rectangles. For example, Figure 5.1 shows the decision node representing the selection of an adjuvant treatment in our simplified ALND model. The two available treatment options are tamoxifen (T) and chemotherapy followed by tamoxifen (CT + T), each indicated by a separate branch. A decision node is terminal when no downstream decisions are explicitly included in the model. To the right of each of the options stemming from a terminal decision node are the leaves, representing the *value* of each option. This is a quantitative summary of the consequences of choosing that option, given all the previous choices and events that led to that leaf. Quantification is typically based on utility theory, as discussed in earlier chapters. For example, in Figure 5.1 the values of the leaves are the expected quality-adjusted survival times associated with T and CT + T. In Section 5.4 we will discuss in detail how these are calculated.

One way of interpreting the expected quality-adjusted survival figures is to think of them as averages of survival times over cohorts of women receiving the same adjuvant treatment, sharing the same covariates (age, ER status and tumor size), and having the same nodal status. In our simplified ALND model we consider three possibilities for the latter: node-positive women, node-negative women, and women whose nodal status is unknown. Women who undergo ALND may fall into either one the first two cohorts; women who do not choose ALND fall into the third. The numbers in Figure 5.1 refer to the cohort with unknown nodal status. The decision problem of choosing an

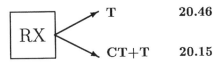

Figure 5.1   Decision node for adjuvant treatment in the simplified ALND decision model. The two branches are alternative adjuvant treatment options: tamoxifen (T) and chemotherapy followed by tamoxifen (CT + T). The figure to the right of each is the corresponding estimated quality-adjusted survival time, assuming no knowledge of nodal involvement.

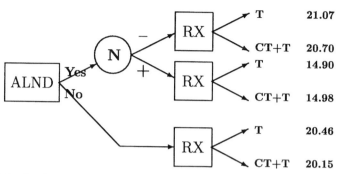

Figure 5.2   Decision tree for the simplified ALND model.

adjuvant therapy for patients in that cohort is solved by looking for the option with the largest expected quality-adjusted length of life. In our example, this option is T.

The three RX decision nodes corresponding to the three cohorts are represented in tree format in Figure 5.2. These nodes are identical in structure but have different estimated quality-adjusted survival times attached to the branches, as a result of the different assumptions made about the nodal status within the cohort in computing the expectations. For example, node-negative women have a longer estimated quality-adjusted survival time than node-positive women.

Knowledge of different patient types, or more generally of different states of the world, is represented in the tree by chance nodes. In our case the relevant chance node is the 'nodal status'. This will be known prior to the decision about adjuvant therapy if an ALND is performed, and will be unknown otherwise. Chance nodes are represented as circles, as exemplified in Figure 5.2. The branches of the chance nodes enter the RX nodes for adjuvant therapy, indicating that the decision will be made having knowledge of that particular outcome. Finally, on the left of Figure 5.2 is the decision node for ALND. The lower branch, corresponding to no ALND, leads directly to the decision about adjuvant therapy, made without information on lymph nodes. The other leads to the axillary lymph node status chance node.

To illustrate, a cohort of, say, 60-year-old patients, ER positive, with a primary tumor of size 0.75 cm enters the ALND node and is divided hypothetically into women who receive ALND and women who do not. The cohort receiving ALND is then divided into positives and negatives according to the appropriate proportions for women with those covariates. In our case 0.0978, or about 10%, test positive. This generates three subcohorts: node-positive women, node-negative women, and women whose nodal status is unknown. Finally, each subcohort is hypothetically subdivided into women who receive T and CT + T. This results in six scenarios, for each of which we computed expected QALYs.

5.3.2 Backward induction

Using these calculations, we can now find a solution using 'backward induction' (Bellman, 1957), a general approach for the solution of multistage decision problems. In our example, the stages are: whether to undergo ALND; and which adjuvant therapy to adopt. Because the solution to the second stage depends on both the decision made at the first stage, and knowledge that may become available between the two stages, it is natural to solve the second stage first, and work our way back toward the earlier stage. Figure 5.3 illustrates backward induction applied to the ALND tree. We start by solving the three terminal problems on the right. On each, we eliminate, or prune, the option with the smaller figure for expected QALYs (shown by two vertical bars). After this elimination, we see that the 'no ALND' branch leads to 20.46 QALYs, while the ALND branch leads to 21.07 for 90.22% of women and 14.98 for 9.78% of women. Computing a weighted average, we obtain the following expected QALYs for the ALND branch:

$$21.07 \times 0.9022 + 14.98 \times 0.0978 = 20.47.$$

This figure is now used as the value of the ALND branch in a reduced tree that consists only of the ALND node and its branches. The value of the ALND branch is larger than that of the no ALND branch, which is then eliminated.

The solution to a multistage decision problem expressed as a tree is the same tree from which all the suboptimal branches stemming from decision nodes have been pruned. At each decision node this pruned tree provides optimal

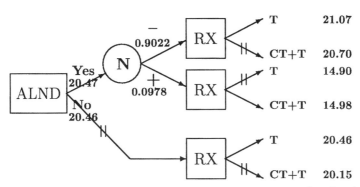

Figure 5.3 Backward induction solution to the decision tree for the simplified ALND model. Vertical bars indicate suboptimal branches that are pruned. Figures are rounded to the fourth significant digit.

decisions that depend on the complete past history of decisions and results of chance nodes. General discussions of this technique, which is a special case of dynamic programming (Bellman, 1957) are found in French (1986), Bernardo and Smith (1994, Chapter 2) and Lindley (1985). Medical applications are presented by Sox $et\ al.$, (1988). A comprehensive and up-to-date overview of available software for computing backward induction solutions to decision trees is provided by the Decision Analysis Society's (2001) site on the World Wide Web.

5.3.3 Measuring the value of prognostic information

Backward induction provides a general tool for iteratively solving complex trees and developing decision strategies that adapt to past decisions and events. In this case, however, there is a perhaps more direct way of understanding why ALND is the preferred option for this cohort. The best adjuvant therapy if ALND is not performed is T. If ALND is performed, the best option is again T if the outcome is negative, but is CT + T if the outcome is positive. So the average QALYs for the positive portion of the ALND cohort, about 10% of all women with the given covariates, are increased by $14.98 - 14.90 = 0.08$. This is the difference between the expected QALYs with and without ALND, assuming that, in either case, women are treated with the optimal adjuvant therapy. For node-negative women, there is no difference in expected QALYs, because the optimal optimal adjuvant therapy is the same in the ALND and no ALND arms. Overall, therefore, ALND has value for this cohort, and can be recommended. Its value is quantifiable as 0.08 QALYs for about 10% of the cohort. The product of these two quantities, that is $0.08 \times 0.1 \approx 0.01$, accounts for the difference observed between the two arms in the dynamic programming solution of Section 5.3.2.

To formalize, let ϕ be the fraction of women testing positive in our cohort. In our example, $\phi = 0.0978$. Also, let q_{C0}, q_{T0}, q_{C1}, and q_{T1} be the QALYs associated with each combination of adjuvant therapy (CT or T) and nodal status (0 for no involvement, 1 for involvement). For women of unknown nodal status, the expected QALYs associated with each combination of adjuvant therapy are called q_C and q_T, and are computed as a weighted average of the QALYs in the two nodal status subgroups. This is because the cohort of women in the no ALND branch is identical to the cohort in the ALND branch, except for what is known about their lymph nodes. Formally,

$$q_C = (1 - \phi)\, q_{C0} + \phi\, q_{C1},$$
$$q_T = (1 - \phi)\, q_{T0} + \phi\, q_{T1}.$$

For example, for the T branch,

$$q_T = 21.07 \times 0.9022 + 14.90 \times 0.0978 = 20.46.$$

In this instance, the value of the noALND branch in the backward induction algorithm is

$$v_N = \max\{q_C, q_T\} = q_T,$$

while the value of the ALND branch is

$$v_A = (1 - \phi)\, q_{T0} + \phi\, q_{C1}.$$

The difference between the two, after some algebraic cancellations, is:

$$\Delta = v_A - v_N = \phi(q_{C1} - q_{T1}). \tag{5.1}$$

This quantity captures the value of ALND as an aid in subsequent clinical decisions, measuring it in units that are of direct clinical interest. While the logic used to derive expression (5.1) is general, the simplicity of form depends on the specific conditions of this example. For example, in the analysis of Section 5.4, expression (5.1) will not necessarily apply. First, we will consider a range of cohorts, for some of which the adjuvant therapy with largest QALYs is the same irrespective of nodal status. For those cohorts, ALND will offer no additional insight in the choice of adjuvant therapy. Even for the cohort considered here, the model will incorporate a direct effect of ALND on both quality of life and survival.

Also, expression (5.1) depends on the rank ordering of the leaves in the specific example of this section. We can write a more general expression for $v_A - v_N$ by noting that

$$v_A = (1 - \phi) \max\{q_{C0}, q_{T0}\} + \phi \max\{q_{C1}, q_{T1}\},$$

and by replacing the expressions for q_C and q_T into the definition of v_N to obtain

$$v_N = \max\{(1 - \phi)q_{C0} + \phi q_{C1}, (1 - \phi)\, q_{T0} + \phi q_{T1}\}.$$

This reveals that v_A and v_N differ only in the order in which the maximization with respect to adjuvant therapy and the expectation with respect to nodal status are taken. This reflects mathematically the fact that v_A and v_N are the values of the best available strategies assuming the information about nodal status is available before the adjuvant therapy decision, and is not available, respectively. When the goal is to maximize length of life, or some quantifiable measure of well-being, information about a patient is valuable only if it may result in changes in subsequent clinical management. When T is superior, or inferior, to CT + T in both node-negative and node-positive women, v_A and v_N coincide.

In our analysis thus far, we have taken the qs and ϕ as a given. In reality these figures are estimates, and there is uncertainty about their actual values. As we will see in more detail in Section 5.4.6, the effect of this uncertainty on the quantification of the value of ALND is an important question.

5.3.4 The value of information

Before studying the ALND problem in more detail, we point out, in the following short and somewhat technical digression, that the idea of measuring the information provided by an observation (in this case the results of the ALND) using a two-stage model in which the first decision is whether to experiment, and the second is how to use the experimental evidence acquired, can be applied generally. Reformulating the ALND problem in more abstract terms, at the first stage a DM decides whether or not to perform a candidate experiment, with unknown outcome x. At the second stage the DM will solve a decision problem by choosing an action a, possibly making use of the experimental outcome x. Before observing x, knowledge of θ is summarized by the prior distribution $p(\theta)$; after observing x, knowledge about θ will be summarized by the posterior distribution $p_x = p(\theta|x)$. The expected utility of action a depends on the current distribution of θ from equation (2.3). We have not brought out this dependence explicitly in the notation, but we need to keep tract of it now, because what is currently known about θ is critical in determining the value of additional experimentation. So the expected utility of action a will be $\mathcal{U}_p(a)$ if the expected utility is computed based on the prior, and $\mathcal{U}_{p_x}(a)$ if it is computed based on the posterior.

The two-stage decision problem above is most effectively solved in reversed order, or by backward induction (DeGroot, 1970), as we did in the ALND tree. The DM first determines the Bayes action $a^*(x)$ and its expected utility $\mathcal{U}_{p_x}(a^*(x))$ as a function of the experimental outcome, for every possible outcome, and then steps back and compares the two options 'experiment' and 'no experiment'. See Section 2.2.2 for more on Bayes actions.

Specifically, under the 'no experiment' arm, the DM would choose action a^* immediately, with expected utility

$$V(p) = \sup_a \mathcal{U}_p(a). \tag{5.2}$$

Under the 'experiment' arm, if $a^*(x) = a^*$ for every outcome x, the experiment does not contribute to the solution of the decision problem to hand. Within a specific decision setting an experiment has value only if it may affect subsequent decisions. How large this value is depends on the utilities assigned to the consequences, and the frequency with which the observation leads to a change of decision. Formally, under 'experiment' the expected utility, averaged with respect to the states of the world and the as yet unobserved experimental outcomes, is

$$\int_{\mathcal{X}} \sup_a \mathcal{U}_{p_x}(a)dx = \int_{\mathcal{X}} V(p_x)dx. \tag{5.3}$$

The ALND model exemplifies the general fact that information can be measured as the difference between the expectation of a maximum and the maximum of an expectation (Ramsey, 1990; Raiffa and Schleifer, 1961; DeGroot, 1984).

The difference

$$V = \int_{\mathcal{X}} V(p_x)dx - V(p)$$

is the amount of utility the DM expects to gain from the experimental outcome. V provides a pragmatic, decision-based way of measuring the value of information. It is specific to the problem and possibly the DM. It adapts to a context, and it is clearly interpretable within that context.

If the experiment comes at a cost, measured by c in the utility scale, it is optimal to perform it whenever $V > c$. This logic can be extended to an arbitrary number of experimental stages (DeGroot, 1970), although computational obstacles arise rapidly as the number of stages increases. The function V is convex, that is,

$$V(p) = V\left(\int_{\mathcal{X}} p_x dx\right) \leq \int_{\mathcal{X}} V(p_x)dx.$$

This ensures that, in expectation, experimentation can never decrease expected utility.

5.4 A more complete decision tree for ALND

5.4.1 Tree structure

We now extend our analysis to a more detailed version of the model. We consider four alternatives for adjuvant therapy, four categories of axillary lymph node involvement, and a direct effect of ALND on survival. Our analysis of the value of ALND will reflect both the informational and the direct therapeutic value.

The alternative of not treating the axilla with either ALND or XRT is not studied in our model. We focused on strategies that provide similarly high probabilities of local control of axillary disease as well as any survival benefits that may accrue from direct treatment of the axilla.

The four alternatives for adjuvant therapy are tamoxifen (T); combination chemotherapy (CT); combination chemotherapy followed by tamoxifen (CT + T); and no adjuvant systemic therapy (NORX). All suitable adjuvant combination chemotherapy regimens are assumed to be equivalent, in terms of both long-term outcome and effects on quality of life. In practice, differences may exist but are likely to be small, and it is unlikely that the relative risk reduction would be affected by nodal status.

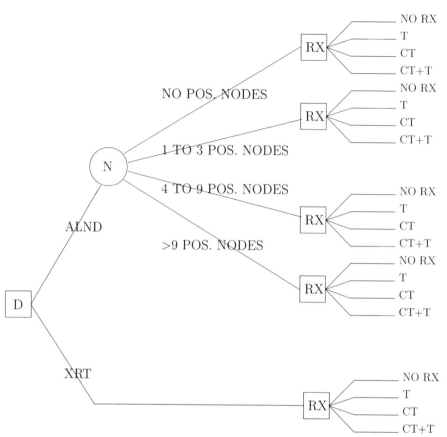

Figure 5.4 Decision tree for the ALND versus XRT decision model. The square on the left represents the decision between ALND and XRT. The circle, N (Nodal status), is an information node, with branches indicating outcomes of the dissection. The RX squares represent decisions about adjuvant therapy.

5.4.2   Population model

Our population model requires a probability distribution of nodal involvement categories by covariate profiles for clinically negative women, and a probability distribution of survival times by nodal involvement category, covariate profile, adjuvant therapy, and ALND status. Although several sources of information about these distributions may be found in the literature, we do not always attempt to formally combine all that is available, but rather utilize the most reliable available source for each distribution. A practical approach is to seek highly generalizable registry or cohort data for describing the natural history of disease under standard treatment, and meta-analyses for describing effects of treatments.

The probability distribution of the number of positive nodes for given tumor size, age, and ER status is based on two components. First we estimate probabilities of nodal involvement by category using data from the population-based Surveillance, Epidemiology, and End Results (SEER) cancer registries (National Cancer Institute, 1997), considering both clinically negative and clinically positive cases. Because our model refers to clinically negative women only, an adjustment is needed. Recht and Houlihan (1995) estimate the probabilities of true negative and false negative physical examinations. This additional information permits adjustment using Bayes' rule. Denote by N the nodal involvement category, by C^- the negative outcome of the clinical examination, and by x the vector of covariates, that is, tumor size, age, and ER status. We have available $p(N|x)$ from SEER and $p(C^-|N)$ from Recht and Houlihan (1995). We need $p(N|x, C^-)$. Assuming that $p(C^-|N)$ is the same for all covariate profiles, we can derive $p(N|x, C^-)$ for each N by evaluating

$$p(N|x, C^-) = \frac{p(N|x)p(C^-|N)}{\sum_n p(N = n|x)p(C^-|N = n)}.$$

Survival distributions were also derived by combining several elements. The survival of patients with positive nodes receiving chemotherapy was estimated using a multivariate Cox regression model with age, ER status, primary tumor size, and nodal involvement as prognostic variables. Patient level data for this purpose were drawn from the combined Cancer and Lenkemia Group B (CALGB) studies 7581, 8082, and 8541 (Wood et al., 1985; Perloff et al., 1996; Wood et al., 1994). The maximum follow-up in these CALGB studies is 19 years. Rates beyond 19 years were extrapolated based on a constant hazard rate (set to the average hazard rate in years 15–19). The hazard rate in the first 19 years is not constant, accounting for greater hazard in the first few months following adjuvant treatment. For this reason our model is more general than standard Markov decision models.

CALGB trials involved only node-positive women. To derive a survival curve for node-negative women, the following strategy was followed. For node-negative women with primary tumors larger than 1 cm, results from the EBCTCG (1992) overview were used to estimate an adjustment ratio between node-positive and node-negative patients receiving chemotherapy. This ratio is 0.42 for the first 4 years and 0.61 for the remainder. For node-negative women with primary tumor smaller than 1 cm, the results of Rosen et al. (1989) were used to derive an adjustment. Mortality from causes other than breast cancer was taken from life tables (Thernau, 1996).

Survival curves of women undergoing adjuvant therapies other than chemotherapy were derived from the previous analysis, together with the results of EBCTCG (1992; 1998a; 1998b) overviews. The results among node-positive patients receiving chemotherapy were used as the baseline and then adjusted in accordance with the risk ratios reported in Table 5.1, to generate survival curves for all treatment options. Effects that were deemed statistically

Table 5.1 Effects of adjuvant treatments (%). Entries are mortality reductions for alternative adjuvant therapies, compared to no adjuvant treatments. Values in parentheses are the corresponding standard deviations. Chemotherapy followed by tamoxifen is not considered for ER negative patients less than 50 years of age.

	Age < 50		Age ≥ 50	
	ER+	ER−	ER+	ER−
NORX	0	0	0	0
T	26 (4.5)	0	32 (10)	0
CT	20 (10)	35 (9)	9 (5)	17 (6)
CT + T	45 (15)	–	40 (11)	11 (4)

unstable (EBCTCG, 1992; 1998a) were not included in the model. Specifically, we assumed that tamoxifen has no effect in premenopausal ER negative patients, and that none of the effects depends on nodal status. Recent evidence from the NSABP B-14 trial (Fisher *et al.*, 1996) also shows a risk reduction for node-negative patients that is comparable with that of previously observed node-positive patients.

Patients receiving ALND also enjoy a 4% reduction in risk of death compared to patients receiving XRT, with standard deviation 4% (EBCTCG, 1992). We do not explicitly consider metastatic disease. This is reasonable if the time and quality of life spent with metastatic disease are not adversely affected by the adjuvant therapy.

5.4.3   Quality of life

Decisions about adjuvant therapy depend on survival benefit as well as on loss of quality of life (QOL) resulting from the therapy itself. We incorporate QOL using hypothetical utility coefficients to provide a quantitative assessment of the impact of various health states on QOL (Pliskin *et al.*, 1980). A utility of 1 corresponds to full health, while 0 corresponds to death. The impact of adjuvant therapy, ALND, and XRT on QOL may vary substantially from patient to patient. We did not elicit utility scores from patients, but instead choose plausible values to represent four different QOL scenarios, based on experience with patients as well as published estimates (Gelber *et al.*, 1991). QOL adjustments are somewhat arbitrary and hard to quantify, but are nevertheless important. If there were no QOL trade-off, all patients with invasive breast cancer would receive chemotherapy and most would also receive tamoxifen. Clearly, physicians and patients presently incorporate QOL considerations in their decision making. Our model is a step toward quantifying this process.

Table 5.2 Illustrative quality adjustment factors, by treatment and month since the beginning of treatment. Adjustment for either ALND or XRT is applied for each subsequent year over and above any adjustment for adjuvant systemic therapy.

Month:	1–6	7–12	13–24	25–60	>60
Q_{mild}					
Tamoxifen	0.99	0.99	0.99	0.99	1
Chemotherapy	0.8	0.95	0.99	0.99	0.99
ALND	0.995	0.995	0.995	0.995	0.995
XRT	0.997	0.997	0.997	0.997	0.997
$Q_{moderate}$					
Tamoxifen	0.99	0.99	0.99	0.99	1
Chemotherapy	0.6	0.9	0.99	0.99	0.99
ALND	0.995	0.995	0.995	0.995	0.995
XRT	0.997	0.997	0.997	0.997	0.997
Q_{severe}					
Tamoxifen	0.8	0.8	0.8	0.8	1
Chemotherapy	0.3	0.6	0.95	0.95	0.95
ALND	0.995	0.995	0.995	0.995	0.995
XRT	0.997	0.997	0.997	0.997	0.997
$Q_{moderate^*}$					
Tamoxifen	0.99	0.99	0.99	0.99	1
Chemotherapy	0.6	0.9	0.99	0.99	0.99
ALND	0.995	0.995	0.995	0.995	0.995
XRT	0.995	0.995	0.995	0.995	0.995

We study four different QOL scenarios, described in Table 5.2. All adjustments for adjuvant therapy vary with time since therapy. Quality adjustments for CT + T (not shown in the table) are obtained using the adjustment factor for CT for the first six months and both adjustments thereafter. In all scenarios, the 'no treatment' option entails no loss of quality, that is, the adjustment is 1. Therefore all quality adjustments are expressed on a scale in which 1 represents the quality of life that a patient would have had prior to recurrence in absence of the treatment considered. This quality of life does not necessarily correspond to that of perfect health.

Our primary analysis uses a scenario representing a moderate impact of adjuvant therapies on QOL, named $Q_{moderate}$. To investigate the sensitivity of our primary results to QOL adjustments, we compared the results obtained under the $Q_{moderate}$ scenario to the results obtained under three additional scenarios, termed Q_{mild}, Q_{severe} and $Q_{moderate^*}$. Q_{mild} exemplifies a patient who experiences little impact on her QOL from her adjuvant therapy (less than in $Q_{moderate}$) and completely recovers from the negative effects of therapy by the end of one year. Q_{severe} reflects a more severe effect of treatment on QOL than $Q_{moderate}$.

There is a QOL impact of both complications from ALND and complications from XRT. Kakuda et al. (1999) conducted a cross-sectional survey of patients being followed by the Breast Surgery Clinic at a university-affiliated urban hospital. They inquired about arm swelling, chest wall pain, decreased mobility, and weakness, and measured upper extremity strength, active range of motion, and circumference. They reported that 70% of patients had at least one complaint, with 18% having moderate to severe symptoms. A significant subset of patients had enduring disability: 21% had notable decrements in strength or range of motion, 9.3% required chronic compression garments for lymphedema, and 6.4% changed their vocational status because of surgical morbidity.

In scenarios $Q_{moderate}$, Q_{mild} and Q_{severe} we assume a negative impact on QOL from ALND greater than that from XRT. The quality adjustment for ALND is 0.995, while that for XRT is 0.997, again based on clinicians' judgment. Both quality adjustments are assumed to be constant over time and are applied over and above any adjustments resulting from adjuvant systemic therapy. In scenario $Q_{moderate^*}$, we explore the implications of using the same QOL adjustment of 0.995 for the negative impact of both ALND and XRT.

To compute quality-adjusted life expectancy using adjustments from Table 5.2, we proceed as follows. Let $S(t)$ be the survivor function for all causes of death combined, starting from the time of treatment, and $Q(t)$ be the quality adjustment for the corresponding unit of time. Then

$$QALY = \int_0^\infty S(t)Q(t)dt.$$

This calculation ignores loss of QOL from other diseases as well as metastatic breast cancer. In view of this, we need to interpret the quality adjustment as relative to a typical health history rather than perfect health.

5.4.4 Solving the tree for a specific patient type

We now consider in detail a specific cohort of patients who are 60 years old at diagnosis, and have an ER positive primary tumor of size between 0.5 and 1 cm. All figures in this section refer to this cohort.

To get a sense for the magnitudes involved we begin by considering survival, without quality adjustment. Table 5.3 considers the proportion surviving 10 years after diagnosis, by nodal status and adjuvant treatment. For each category of nodal involvement, the table gives the corresponding probability, and the change in the proportion of 10-year survivors, compared to no

Table 5.3 Proportion surviving after 10 years (increment compared to the no-treatment arm), by adjuvant treatment type and nodal group for the example cohort.

Nodal status	Nodal status probability	Increment in proportion surviving after 10 years		
		T	CT	CT + T
0	0.9	0.016	0.004	0.019
1–3	0.07	0.089	0.024	0.107
4–9	0.02	0.098	0.026	0.118
>9	0.01	0.114	0.029	0.139
Unknown		0.023	0.006	0.028

treatment. For example, the fraction of women with one to three positive nodes after ALND is 7%. If one to three positive nodes were found, the difference in 10-year survival between administration of tamoxifen and no adjuvant therapy would be 8.9 percentage points. Administration of tamoxifen and chemotherapy would lead to a difference of 10.7 percentage points, and chemotherapy alone to a difference of 2.4 percentage points. If ALND is not performed, the additional survival difference can be computed as a weighted average of the survival differences in the nodal involvement categories. These weighted averages are reported in the 'unknown' nodal status row.

A common adjuvant treatment strategy in this case is to administer tamoxifen to node-negative women and to women of unknown nodal status (not undergoing ALND), and to administer tamoxifen plus chemotherapy to node-positive women. Under this strategy, 90% of the women in the cohort would receive the same adjuvant treatment whether or not they elect to have an ALND. The remaining 10% would have tamoxifen if they did not choose ALND, and tamoxifen plus chemotherapy if they did. The improvement in survival resulting from this is 1.8 percentage points for 3% of the women (those with 1–3 nodes involved), 2 percentage points for 2% of the women, and 2.5 percentage points for 1% of the women. On average for the cohort, this results in an increase in survival at 10 years slightly greater than 0.1%. This increase in survival will be termed the 'information' effect of ALND, as opposed to the 'direct' effect it may have on survival via local control of disease. Both effects contribute to the total benefit of ALND.

Decision analysis formally determines the best adjuvant strategies by considering survival comparisons after quality adjustment. Table 5.4 illustrates this, summarizing the average QALYs by adjuvant therapies and nodal status, for scenarios Q_{mild}, $Q_{moderate}$, and Q_{severe}. The table for the $Q_{moderate^*}$ scenario differs only slightly from that of $Q_{moderate}$ and is omitted.

MODELING IN MEDICAL DECISION MAKING

Table 5.4 Expected QALYs for adjuvant therapy arms by nodal group and quality of life in the scenario. This table summarizes the calculations used in solving the decision tree. Boxed values indicate the best treatment option for a given nodal status.

Nodal status	Probability	Additional QALYs by adjuvant RX			
		None	T	CT	$CT + T$
Q_{mild}					
0	0.9	20.39	20.86	20.22	20.64
1–3	0.07	12.58	14.7	12.89	14.91
4–9	0.02	11.58	13.79	11.91	14.04
>9	0.01	9.44	11.72	9.79	12.03
Unknown		19.54	20.2	19.42	20.02
$Q_{moderate}$					
0	0.9	20.39	20.86	20.09	20.51
1–3	0.07	12.58	14.7	12.77	14.79
4–9	0.02	11.58	13.79	11.79	13.92
>9	0.01	9.44	11.72	9.66	11.91
Unknown		19.54	20.2	19.3	19.9
Q_{severe}					
0	0.9	20.39	19.95	19.02	19.24
1–3	0.07	12.58	13.84	11.99	13.76
4–9	0.02	11.58	12.94	11.04	12.92
>9	0.01	9.44	10.89	9.01	11
Unknown		19.54	19.29	18.25	18.65

The boxes highlight the treatment that results in the greatest gains in QALYs. We will refer to these as the best options, for conciseness.

In scenarios Q_{mild} and $Q_{moderate}$, the best adjuvant treatment option is tamoxifen if ALND is not performed (nodal status unknown). If ALND is performed, node-negative patients receive tamoxifen, while node-positive patients receive chemotherapy plus tamoxifen. In scenario Q_{severe} the best option if ALND is not performed is no adjuvant treatment, because of the more severe impact of adjuvant treatment on QOL. If ALND is performed, node-negative patients receive no adjuvant treatment, patients with less than

Table 5.5 Expected QALYs for the ALND and XRT arms by quality of life scenario. Entries in the ALND arm are obtained by averaging the boxed values in Table 5.4 with respect to the nodal involvement category. On the right is the difference ΔQALYs between the two arms.

	Arm	Expected QALYs	ΔQALYs
Q_{mild}	ALND	20.24	0.04
	XRT	20.20	
$Q_{moderate}$	ALND	20.23	0.03
	XRT	20.20	
Q_{severe}	ALND	19.71	0.17
	XRT	19.54	
$Q_{moderate^*}$	ALND	20.23	0.07
	XRT	20.16	

10 positive nodes receive tamoxifen, and patients with more than nine positive nodes receive chemotherapy plus tamoxifen.

For this cohort, in all QOL scenarios ALND can lead to a change in the appropriate adjuvant therapy and therefore has an 'information' effect. Table 5.5 summarizes the expected QALYs for the ALND and XRT arms, and their difference ΔQALYs in the four QOL scenarios. These differences are the result of the information and direct effect, and, in all but the $Q_{moderate^*}$ scenario, the differential loss of quality of life for ALND and XRT. Comparison of $Q_{moderate}$ and $Q_{moderate^*}$ shows that the differential loss of QOL for ALND and XRT subtracts about 0.04 QALYs from the combined information and direct effects of 0.07 QALYs. Q_{severe} is the scenario in which ΔQALYs is largest for this cohort. In this scenario, node-positive women are expected to enjoy in excess of one QALY from the change in adjuvant therapy resulting from ALND, a much larger gain than we have in the other scenarios.

5.4.5 Mapping the value of ALND in the covariates space

Decisions about adjuvant therapy and ALND may vary depending on patient characteristics. Different patients have different probabilities of finding positive nodes. The effectiveness of adjuvant therapies and the importance of the competing causes of death also vary with the patient, the latter phenomenon increasing with age.

Our model incorporates three covariates. It is interesting to explore the effect of changes in these covariates on ΔQALYs. Table 5.6 summarizes ΔQALYs for all patient types in the $Q_{moderate}$ scenario. For 70-year-old patients with small (less than 0.5 cm) ER positive tumors ΔQALYs is -0.01, indicating that ALND has smaller expected QALYs than XRT by about 3 or 4 expected quality-adjusted days of life. For all other patient types ΔQALYs is positive. Differences range from a few days to about 6 months, quality-adjusted. The women most likely to benefit are younger and have tumors larger than 1 cm.

To further explore the results of Table 5.6, Figure 5.5 shows recommended adjuvant therapy choices for all patient types, in the four quality scenarios. At each rectangle, the string of four letters represents the choices of adjuvant therapy by nodal status (0, 1–3, 4–9, >9 from left to right) given that ALND is performed. The rightmost letter represents the choice of adjuvant therapy if ALND is not performed. The shading corresponds to the values of ΔQALYs reported in Table 5.6, with darker colors indicating lower differences.

In many rectangles in the table, the optimal adjuvant therapy is the same irrespective of nodal status. For example, for ER negative women, in most patient types the optimal adjuvant therapy choice is chemotherapy alone, irrespective of nodal status. For those patient types, ALND has no information effect, and the ΔQALYs observed is the balance of the direct effect and the loss of QALYs. If the assumption of a differential direct effect compared to XRT is called into question, the worthiness of ALND for those women is substantially reduced. The only ER negative women for which ALND has an information effect are older women with small tumors who, if node-negative, would not receive chemotherapy.

Table 5.6 ΔQALYs by age, ER status, and tumor size, in the $Q_{moderate}$ scenario. Negative values correspond to patient types for which XRT has higher expected QALYs than ALND.

Tumor size	ER−					ER+				
	Age					Age				
	30	40	50	60	70	30	40	50	60	70
<0.5	0.26	0.18	0.11	0.07	0.04	0.08	0.05	0.03	0.01	−0.01
0.5–1	0.36	0.25	0.15	0.09	0.07	0.19	0.10	0.08	0.03	0.00
1–2	0.50	0.40	0.30	0.20	0.11	0.46	0.35	0.23	0.17	0.07
2–3	0.48	0.39	0.30	0.21	0.12	0.48	0.37	0.25	0.17	0.09
3–4	0.46	0.38	0.30	0.21	0.13	0.49	0.38	0.26	0.17	0.09
4–5	0.43	0.37	0.29	0.21	0.13	0.50	0.39	0.27	0.18	0.11
>5	0.34	0.29	0.25	0.19	0.13	0.47	0.39	0.30	0.20	0.14

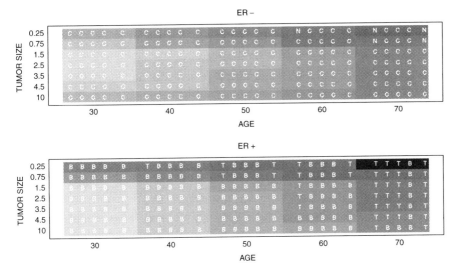

Figure 5.5 Optimal adjuvant therapies by patient types and nodal status: C, chemotherapy; T, tamoxifen; B, both; N, no adjuvant therapy. Each rectangle corresponds to a patient type. The four letters on the left of each rectangle indicate the optimal therapy for each of the four categories of nodal involvement, starting with no involvement on the left. The right letter is the optimal strategy when nodal status is unknown. Shading is proportional to ΔQALYs, reported in Table 5.6.

For ER positive women the information effect of ALND is more common. All receive tamoxifen, while the optimal adjuvant therapy may or may not include the addition of chemotherapy. Again, older women and women with smaller tumors are those in which the optimal adjuvant therapy depends on the nodal status. Interestingly, for younger women the information effect is likely to be smaller, but the direct effect is larger, as it stretches over a longer expected life length. The trade-off between these two effects and the loss of QOL determined the overall shape of the ΔQALYs surface.

Table 5.7 summarizes ΔQALYs for all patient types, in the Q_{mild}, Q_{severe}, and $Q_{moderate^*}$ scenarios. This analysis illustrates that there are patient types in which the preferred choice is sensitive to the QOL adjustments and others in which it is not. For younger women with larger tumors the ΔQALYs are similar across scenarios and depend mostly on the direct effect. The subset of women in which the results are most sensitive to attitudes towards QOL are premenopausal women with small primary tumors. In this group, node-negative women may or may not opt for chemotherapy depending on their personal values. If they expect a small QOL reduction from chemotherapy, ALND is of little or no value to them since they will elect to receive chemotherapy anyway. If they expect a greater QOL reduction, ALND becomes more desirable, as node-negative women may wish not to receive chemotherapy.

Table 5.7 ΔQALYs by age, ER status, and tumor size, in the Q_{mild}, Q_{severe}, and $Q_{moderate*}$ scenarios. Negative values correspond to patient types for which XRT has higher expected QALYs than ALND.

Q_{mild}

Tumor size	ER−					ER+				
			Age					Age		
	30	40	50	60	70	30	40	50	60	70
<0.5	0.26	0.18	0.10	0.06	0.05	0.08	0.05	0.03	0.01	−0.00
0.5–1	0.36	0.24	0.15	0.09	0.04	0.19	0.10	0.09	0.05	0.01
1–2	0.50	0.40	0.30	0.20	0.11	0.46	0.35	0.23	0.14	0.08
2–3	0.48	0.39	0.30	0.21	0.12	0.48	0.36	0.25	0.15	0.10
3–4	0.46	0.38	0.30	0.21	0.13	0.49	0.38	0.26	0.17	0.10
4–5	0.43	0.37	0.29	0.21	0.13	0.50	0.39	0.27	0.18	0.10
>5	0.34	0.29	0.25	0.19	0.13	0.47	0.39	0.30	0.20	0.12

Q_{severe}

Tumor Size	ER−					ER+				
			Age					Age		
	30	40	50	60	70	30	40	50	60	70
<0.5	0.25	0.17	0.14	0.07	0.03	0.21	0.16	0.13	0.08	0.02
0.5–1	0.35	0.24	0.21	0.11	0.05	0.68	0.33	0.11	0.17	0.04
1–2	0.48	0.38	0.29	0.23	0.12	0.45	0.33	0.25	0.15	0.07
2–3	0.46	0.38	0.29	0.25	0.13	0.46	0.35	0.29	0.17	0.09
3–4	0.44	0.37	0.28	0.24	0.13	0.47	0.37	0.33	0.20	0.09
4–5	0.42	0.35	0.28	0.22	0.13	0.48	0.37	0.35	0.22	0.11
>5	0.33	0.28	0.24	0.19	0.13	0.45	0.37	0.29	0.22	0.15

$Q_{moderate}$

Tumor Size	ER−					ER+				
			Age					Age		
	30	40	50	60	70	30	40	50	60	70
<0.5	0.34	0.24	0.16	0.11	0.07	0.17	0.12	0.09	0.05	0.02
0.5–1	0.42	0.30	0.20	0.12	0.10	0.27	0.17	0.13	0.07	0.03
1–2	0.53	0.44	0.33	0.23	0.13	0.52	0.40	0.27	0.20	0.10
2–3	0.51	0.42	0.33	0.23	0.14	0.53	0.41	0.29	0.20	0.11
3–4	0.49	0.41	0.32	0.23	0.14	0.54	0.42	0.30	0.20	0.12
4–5	0.46	0.39	0.31	0.23	0.15	0.54	0.43	0.31	0.21	0.13
>5	0.36	0.31	0.26	0.21	0.14	0.51	0.42	0.33	0.23	0.16

The largest values of ΔQALYs occur in the severe scenario, because of the strong additional contribution of the information effect.

5.4.6 Uncertainty and probabilistic sensitivity analysis

Our analysis so far has considered all input parameters as fixed. In reality, these input are estimates and there is uncertainty about their actual values. In general, in a decision analysis, we can distinguish between at least two sources of uncertainty: one is the uncertainty about the individual history of a woman as compared to that of a cohort of similar women. The other is the uncertainty about the cohort itself, which is known only up to statistical error. In a Bayesian approach, both uncertainties can be quantified by probability distributions. The distribution of the individual histories conditional on the parameters represents variation within the cohort; the distribution of the parameters represents the plausibility of alternative potential cohort descriptions.

Indicate by θ the vector of all the unknown parameters. Uncertainty can be quantified by assigning to θ a distribution reflecting what is known about its possible values. In a Bayesian framework, this distribution is provided by posterior distribution from the statistical analysis of source data, so that the connection between statistical estimation and decision making is seamless. We assume that the estimates used in the previous section are the expectations of this distribution. The quantity of interest Δ is a function of the vector θ. Uncertainly about individual histories is averaged out in computing Δ, for a fixed vector θ.

Because θ is random, Δ is also random. Quantifying the uncertainty in Δ is interesting because it provides a way of measuring how reliably we can claim that ALND is indicated for a cohort. This quantification is an example of probabilistic sensitivity analysis, discussed in Section 3.4.1. Even in relatively simple problems such as ours, it is most conveniently done by simulation. Parameter values are drawn from their distributions; Δ and other quantities of interest are calculated for each draw; the resulting distributions are studied. Probabilistic sensitivity analysis by simulation has been advocated by various authors, including Doubilet et al. (1985) and Critchfield and Willard (1986) in nonsequential decisions. Parmigiani (2001) describes additional complexities arising in multistage decision problems.

In this section, we illustrate a probabilistic sensitivity analysis for the whole model. Real-valued parameters were assigned normal distributions, while probabilities and percentage of risk reduction were assigned beta distributions. Means were set at the baseline value of the model, while variability was based on the standard deviations indicated in our previous discussion, which in turn are based either on the reported standard deviations in the original studies, or on the estimates obtained from source data. All parameters estimated using SEER data were handled using a Dirichlet posterior distribution obtained from uniform priors – for details, see Gelman et al. (1995). For the

Cox proportional hazard model, the baseline hazard is fixed, and only the co-efficients are drawn. There are a total of 128 parameters related to the nodal involvement probabilities and 11 parameters related to the prognostic model that are being drawn at random. Distributions of parameters within the SEER analysis and within the Cox proportional hazards model are dependent, but the two sets are independent of each other and of the risk ratios.

We focus again on the cohort of Section 5.4.4, that is, patients who are 60 years old at diagnosis, and have an ER positive primary tumor of size between 0.5 and 1 cm. We study the distribution of two quantities: ΔQALYs and Δ^*QALYs. They both represent the difference between the average QALYs in the ALND and XRT cohorts, with one critical difference. ΔQALYs is calculated by assuming that the adjuvant therapy choices are those of Table 5.4, while Δ^*QALYs is calculated by rederiving the optimal adjuvant therapy choice for each draw of simulated parameters. In a two-stage decision such as this, ignorance of the correct value of the model parameters implies ignorance of the correct decision at the second stage. This translates into uncertainty about the decision at the first stage as well. ΔQALYs incorporates this uncertainty, and realistically models the actual state of knowledge, as parameter values will not be known when the adjuvant therapy decision needs to be made.

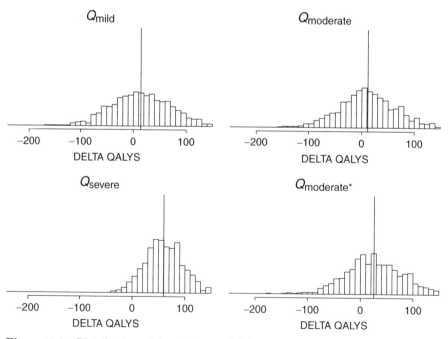

Figure 5.6   Distribution of the difference ΔQALYs, in days, by QOL scenario, after 1600 simulations of input parameters. Vertical bars indicate means of the simulated values.

Figure 5.6 shows histograms of the distribution of ΔQALYs, by QOL scenario, based on 900 simulated values of the model parameters. Vertical bars indicate means of the simulated values. Importantly, the estimate of Δ based on 'plugging in' the estimated vector θ is not necessarily the same as the expectation of the distribution of Δ, even though this may be true for individual components of θ.

In the mild and moderate scenarios, the variation in the ΔQALYs histograms, attributable to lack of knowledge about the model parameters, far outweighs the mean value of ΔQALYs. In the severe scenario, however, ΔQALYs is almost always positive. This stresses the importance of adding an uncertainty analysis to the computations of Section 5.4.4. For some patient types and QOL preferences, conclusions about ALND can be reached with confidence, while for others odds are about even that the procedure is of value.

In a probabilistic sensitivity analysis of a multistage problem with hundreds of input parameters there are endless interesting analyses that can be done using the simulation output. In general, three critical quantities are the following:

$PC = P(\Delta = \Delta^*)$ The probability that the adjuvant therapy choice at the second stage is correct under the true value of the parameters.

$PD = P(\Delta^* > 0)$ The probability that ALND is of value under the true value of the parameters. When there are no direct effects, this is the probability that ALND is of diagnostic value.

$PE = P(\Delta > 0)$ The probability that ALND is ethical, that is, it does more good than harm given the current state of knowledge and associated optimal adjuvant strategy.

Table 5.8 reports PC, PD and PE in the moderate and severe QOL scenarios, based on 900 simulated values. In both cases there is substantial uncertainty as to the correct adjuvant therapy, with the best procedure. In the moderate scenario there is also considerable uncertainty as to whether the ALND arm will do better than the XRT arm under the chosen strategy, with less than

Table 5.8 Values of PC, PD and PE in the moderate and severe QOL scenarios, based on 900 simulated values.

	Q_{moderate}	Q_{severe}
PC	0.29	0.03
PD	0.75	0.97
PE	0.59	0.96

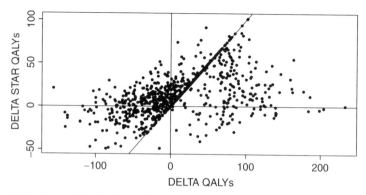

Figure 5.7 Joint distribution of ΔQALYs and Δ*QALYs, in days, after 900 simulations of input parameters. The diagonal line is ΔQALYs = Δ*QALYs.

60% of simulated parameter sets leading to a better outcome. ALND has some diagnostic value in 75% of simulated values, indicating that better knowledge of the parameters could have a high impact via improved adjuvant therapy choices. By contrast in the severe scenario, both PD and PE are very high and close to each other, despite the very low value of PC. This indicates that while the optimal adjuvant choice is very likely to be incorrect, only small variations from this strategy occur as we draw different parameter values. ALND can be recommended with confidence in this QOL scenario.

The determination of PC, PD and PE is examined further in Figure 5.7, graphing the joint distribution of ΔQALYs and Δ*QALYs. Each point corresponds to a simulated set of parameter values. Points on the diagonal line are such that ΔQALYs = Δ*QALYs, corresponding to draws for which the adjuvant therapy choices of Table 5.4 are correct. The proportion of points on the line is PC. The fraction of points in the two top quadrants is PD and that in the two right-hand quadrants is PE. Points to the right of the diagonal line correspond to parameter values for which ALND is of greater value under the current optimal strategy than it is under the optimal strategy for the given θ. For example, it is possible that the optimal strategy for the given θ may be to administer tamoxifen plus chemotherapy to all nodal status categories, and therefore to women in the XRT arm as well. This means that Δ* has no information component. It is still possible, however, that the strategy of Table 5.4 (tamoxifen to node-negative and XRT women, and tamoxifen plus chemotherapy to all others) may have an information advantage. In this case, by administering chemotherapy to all women, while reducing the difference between the two arms, we would increase QALYs in both.

Another important feature of probabilistic sensitivity analysis is the ability to assess the impact of individual parameter uncertainty on the conclusions. The general idea is illustrated in Figure 5.8, graphing ΔQALYs in the moderate scenario versus one of the critical parameters of the model: the

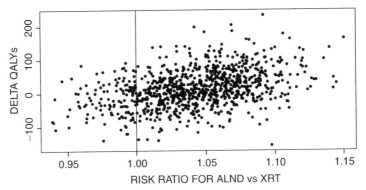

Figure 5.8 Joint distribution of ΔQALYs and the risk ratio for ALND versus XRT. The vertical line corresponds to a risk ratio of one, or the same effect on mortality from both procedures. Risk ratios greater than one indicate that the risk reduction from ALND is greater than that from XRT.

risk ratio for ALND versus XRT. This parameter captures the relative effect of ALND versus XRT on overall survival as a result of local control. A value of 1 indicates identical effects, while values greater than 1 indicate that the risk reduction from ALND is greater than that from XRT. The mean value of this parameter is 1.042 and the standard deviation used in the probabilistic sensitivity analysis is 0.0369. The effect of increasing the risk ratio on ΔQALYs is clear. Because the relationship between the risk ratio and ΔQALYs is roughly linear, we can measure the proportion of the variation in ΔQALYs that is attributable to variations in the risk ratio by computing R^2. In this example R^2 is 14.4%, a substantial proportion considering the large number of parameters entering the model. The literature on analyzing the output of computer experiments and sensitivity analyses is extensive. Reviews and approaches for multiple parameters are presented by Kleijnen and Helton (1999) and Kleijnen (1997).

6

Chronic disease modeling

6.1 Summary

In this chapter we present a decision model exemplifying the type of multicomponent approaches needed to address complex decisions in chromic disease processes. The context is that of studying the appropriate frequency of examinations in screening for breast cancer, which involves substantial uncertainty and difficult trade-offs. We discuss a general plan and initial work towards a comprehensive approach. The current implementation of the model presented here has important limitations, and some of its components are work in progress. The general plan, however, should give some perspective on both the strengths and the challenges of an integrated approach to decision making, and could help chart the way for future more complete implementations. This case study draws on material from Chapter 5, as well as several earlier articles, including Parmigiani and Kamlet (1993), Parmigiani (1997; 1998), and Parmigiani and Skates (2001).

6.2 Model overview

6.2.1 Modeling in breast cancer screening

Mammography is a radiological procedure for the detection of breast cancer in asymptomatic patients. Its purpose is to identify breast cancer at an early stage of development, in the hope of a more effective treatment. Mammographic screening is considered a critical component of the control of breast cancer and is widely recommended, primarily on the basis of evidence of mortality reduction from large-scale randomized trials (Shapiro *et al.*, 1988; Morrison, 1989). Despite extensive study, there remains uncertainty as to who can benefit from screening, and the extent of that benefit. This is because the benefits of screening depend on complex interaction among several factors, including the ability of a mammogram to detect cancer sufficiently early; the time window during which such detection is possible, and its relation to the interval between screening examinations; the relative advantage of early detection compared to standard detection at the onset of symptoms; the age distribution of onset of presymptomatic cancer; competing causes of mortality; and others.

Screening recommendations available to clinicians differ with regard to screening modalities, and different variants have the potential for a large public health and economic impact. For example, controversy surrounds the question of whether to recommend regular mammograms to women in their forties. There is disagreement over the interpretation of the results of randomized clinical trials, as evidence of benefit varies across trials (Berry, 1998). Even if one accepts that the screening arms in the trials did better than the corresponding control arms, there is controversy over the relevance of that result for individual women's decisions. Task forces in both the United States (Gordis et al., 1997) and Canada (Canadian Task Force on Preventive Health Care, 2001), assessed the evidence but did not find it sufficiently strong to make general recommendations, emphasizing that 'women should be informed of the potential benefits and risks of screening mammography and assisted in deciding at what age they wish to initiate the manoeuvre' (Canadian Task Force on Preventive Health Care, 2001). Other difficult issues are the appropriate frequency of screening examinations; whether women who are at increased risk of breast cancer would benefit from more frequent screening; whether it is cost-effective to provide insurance coverage for screening mammograms; and what the impact of new technology would be.

In this scenario, a decision model can provide guidance by bringing together and integrating evidence from several sources, including the epidemiology and genetics of risk factors, relevant clinical trials of secondary prevention and treatment, and knowledge about tumor growth mechanisms. A decision model can also help give a realistic assessment of uncertainty about the relative merits of alternative choices, and provide the basis for a cost-effectiveness analysis of specific interventions, or the maximization of specific utility functions. Interest in model-based approaches to addressing issues in cancer screening dates back at least to the 1970s. Early work includes Knox (1973), Schwartz (1978), Eddy (1980), and Habbema et al. (1985). Van Oortmarssen et al. (1995) provide a recent review of modeling issues.

6.2.2 Model components

We begin with a broad-brush description of the general plan of the model. It is useful to think of it as the combination of four components:

1. A natural history model, describing the progress of breast cancer in the absence of screening, from the onset of detectable but asymptomatic disease, through tumor growth, to symptoms.

2. A sensitivity model, describing the probability of early detection given tumor and patient characteristics at the time of the examination.

3. A nodal involvement model, describing the probability distribution of the number of axillary nodes involved with cancer, given age and tumor size at detection.

4. A survival model describing length and quality of life after diagnosis of breast cancer, as a function of tumor characteristics at the time of detection. This component will be the same as that used in Chapter 5 for analyzing axillary lymph node dissection.

Figure 6.1 provides an influence diagram (Forrester, 1961; Clemen and Reilly, 2001) representing the interactions among these components.

The potential benefits of screening derive from the ability to detect the disease at an early stage. Screening can lead to early discovery of cancer, when the tumor is small and unlikely to have developed metastases outside the breast. Small, nonmetastatic tumors can be treated more effectively and less aggressively, with better survival and QOL. On the other hand, screening leads to the discovery of cancer that would have not otherwise affected a woman's health. This represents a considerable loss of QOL. Also, early detection can prolong the portion of one's life spent as a cancer survivor.

In a decision model, we can analyze these complex interactions and trade-offs by simulating individual natural histories, inclusive of tumor growth rate,

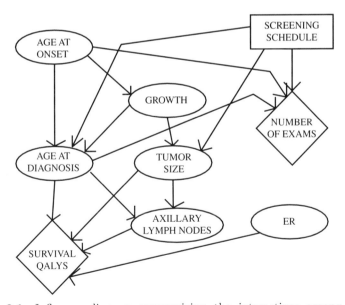

Figure 6.1 Influence diagram summarizing the interactions among the key variables in the decision model. Ovals represent random variables, rectangles represent decision variables, in this case the screening strategy as a whole, and diamonds represent outcomes of interest. Arrows indicate direct effects in the specification of conditional probability distributions, and implicitly specify the conditional independence structure of the model.

and then study the consequences of different screening strategies on each individual. For a patient with preclinical disease, the natural history model provides a way of simulating age of onset and sojourn time, or, equivalently, average growth rate. Based on these it is possible to simulate the age of the woman and the tumor size at the time of diagnosis. These variables can then be used to predict the woman's survival and quality-adjusted survival.

Model components contribute to this plan in the following way. The joint distribution of the age of onset and tumor growth variables is determined by the natural history model of Section 6.3. For any given screening strategy and sensitivity model, these determine a distribution of age at diagnosis and tumor size at diagnosis. The sensitivity model provides the probability of a positive mammogram given age at examination and tumor size at examination. The nodal involvement model provides the conditional distribution of the number of axillary lymph nodes displaying metastatic cancer given age at diagnosis and tumor size at diagnosis. Finally, the survival model gives the conditional distribution of length of life given the prognostic factors age at diagnosis, tumor size at diagnosis, number of nodes involved, and estrogen receptor status. The QALY variable refers to the quality-adjusted survival. The number of examinations is a surrogate for the costs of screening. Here we do not explicitly consider the downstream cost implications on treatment and long-term care.

6.3 Natural history model

6.3.1 States and transitions

The core of our scheme is the natural history model, which represents the health histories of a cohort of women at risk for breast cancer, when no screening takes place. As in the μSPPM of Section 1.5.1, we construct the natural history model by specifying states, transitions, and transition probabilities. Relevant disease states for screening are: H, women who are either healthy or have breast cancer at a stage that is too early to detect by mammography; P, women who have detectable preclinical breast cancer, that is, breast cancer that could be found by mammography but has not yet led to symptoms; C, women who have been diagnosed clinically with breast cancer; and D, women who have died.

This set of states is the minimal set that is required to model early detection of occult disease. Pioneering work of Zelen and Feinleib (1969) considered the sequence H $\to$ P $\to$ C. Explicit consideration of competing causes of mortality was introduced later, for example in microsimulation models (Schwartz, 1978; Habbema et al., 1985) and in models for the study of optimal screening for occult disease (Tsodikov and Yakovlev, 1991; Parmigiani, 1993; Yakovlev and Tsodikov, 1996). The definition of state P is specific to the type of screening technique used. For example, two mammographic techniques with different

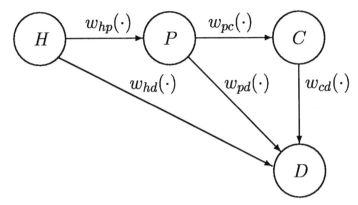

Figure 6.2 Summary of states, possible transitions, and transition densities for the natural history model. This scheme describes the progress of breast cancer in the absence of screening. All instantaneous probabilities of transition are indicated next to the respective transition. The two subscripts correspond to the origin and destination states respectively.

minimum detectable size would imply different definitions for state P. One way to incorporate them jointly into a single natural history model would then be to define P as the state in which the tumor is detectable by at least one mode of screening, and use the tumor growth variable and the sensitivity model to capture detectability with the other modality.

Figure 6.2 summarizes states and transitions: states are represented by circles, and possible transitions by arrows. Transitions from H to C without entering P are not considered, although it is possible for a woman to spend an arbitrarily small amount of time in state P. Also reverse transitions are excluded, consistent with the progressive nature of the disease. In summary, transitions can occur from H to P, from P to C and from any state to death. Transitions between states can occur at any time.

Additional prognostic variables, such as axillary node involvement and tumor size at detection, are collectively denoted by x, and not represented explicitly in Figure 6.2. These variables are used in other model components to predict survival and capture the effects of early detection by screening. Some of these variables could be included directly in the natural history component by using additional states. For example, we could subdivide state C into more specific states depending on nodal status. For the purposes of our analysis it is simpler to keep track of prognostic variables via the transition probability of the time in C.

In general, there is a trade-off between the complexity of the state representation and the complexity of the transition probability structure. A more detailed state representation makes it more challenging to estimate the transition densities of general forms as done here. For example, the microsimulation model MISCAN (Habbema *et al.*, 1985; van Oortmarssen *et al.*, 1995), considers multiple stages of preclinical disease, which are not considered here, but uses simple parametric transition distributions. Our specification captures differences in stages of preclinical disease, leading in turn to differences in prognosis, via the axillary lymph node involvement model, which depends on tumor size at detection, and the survival model.

6.3.2　*Transition densities and sojourn time distribution*

The natural history model represents histories in continuous time, and over the entire positive real line. This offers a convenient analytical framework for modeling and optimization with respect to examination times or other intervention variables. In our simulation we will consider a discretization using three-month intervals, and a finite horizon, with a maximum life length of 110 years. The model considers a cohort from birth. To draw conclusions for a cohort of women that are asymptomatic at age, say 40, it is sufficient to condition on this event. For example, in the simulation, we would simply focus on simulated histories that are still asymptomatic at age 40, and exclude all others.

Each woman spends a random amount of time in various states. Random variables t and s are the times spent in H and P, respectively. Accordingly, a woman will leave these states at times t and $t + s$. The random variable v denotes the time in state C, if the disease is not detected by screening. If a woman dies while in H, then $s = 0$ and $v = 0$. Likewise, if she dies while in P, then $v = 0$. Transition densities describe the proportion of women making a certain transition per unit of time (or per time interval in the discrete version), out of the total number of women in the initial cohort. For example $w_{hp}(y)$ represents the fraction of women making the transition from H to P in a neighborhood of time y. Notation for all transition densities is summarized in Figure 6.2.

Transition densities are related to the distribution of the time spent in various states. For example, the random time t spend in H is distributed according to the probability density $w_{hp}(y) + w_{hd}(y)$. If there are multiple destinations from a given state, individual transition densities w do not need to integrate to 1 over the real line. However, the sum of all densities exiting a given state, such as $w_{hp} + w_{hd}$, will integrate to 1. Following Zelen and Feinleib (1969), we use the following notation for tail probabilities:

$$W_{hd}(t) = \int_t^\infty w_{hd}(t')dt',$$

$$W_{hp}(t) = \int_t^\infty w_{hp}(t')dt',$$

so that $W_{hp}(0)$ is the probability of developing preclinical disease over the course of one's lifetime. Also, $W_{hp}(0) + W_{hd}(0) = 1$.

Similar considerations and notation apply to the distributions originating from state P, with the addition that we allow for dependence upon the time of the previous transition, to account for the fact that tumor growth is likely to be faster in younger women (Moskowitz, 1986; Spratt $et\ al.$, 1986). Formally, $w_{\mathrm{pc}}(s|t)$ and $w_{\mathrm{pd}}(s|t)$ depend on t. Similarly to previous definitions,

$$W_{\mathrm{pd}}(s|t) = \int_s^\infty w_{\mathrm{pd}}(s'|t)ds',$$

$$W_{\mathrm{pc}}(s|t) = \int_s^\infty w_{\mathrm{pc}}(s'|t)ds'.$$

Here $W_{\mathrm{pd}}(0|t)$ is the probability of developing clinical breast cancer (in absence of screening) conditional on entering the preclinical state at age t. For older women, this probability is far smaller than 1, as a result of competing causes of mortality during the preclinical state. Conversely, $W_{\mathrm{pd}}(0|t)$ represents the probability of dying of causes other than cancer while the disease is still preclinical. As earlier, $W_{\mathrm{pc}}(0|t) + W_{\mathrm{pd}}(0|t) = 1$.

Finally, $w_{\mathrm{cd}}(v|t, s, x)$ is the distribution of the time spent in C. This describes survival after diagnosis, so it can also be considered part of the survival model. It depends on t and s, as well as prognostic variables x at detection. The marginal distribution of the prognostic factors at detection is described by a separate probability model, $p(x|t, s)$. One approach to modeling the efficacy of screening is to study how early detection by screening affects the distributions of x and v.

The distributions described so far are sufficient to determine the probabilities of any path through the network defined by Figure 6.2. An alternative way of representing the same probabilities is by specifying the distributions of the times spent in each state, along with the conditional distributions of the direction of transition given the transition times. This is the approach that was described in Section 1.5.2 for the estimation of multistage models. It is especially useful when all transitions are observable, and direct data are available, which is not the case here.

There is a correspondence between these two ways of specifying the probabilities for the model. For example, the probability of making a transition out of H at time t is $w_{\mathrm{hp}}(t) + w_{\mathrm{hd}}(t)$. Conditional on a transition out of state H at time t, the probability of moving to P is

$$\frac{w_{\mathrm{hp}}(t)}{w_{\mathrm{hp}}(t) + w_{\mathrm{hd}}(t)}.$$

When estimating transitions in competing risk models, we will focus on the conditional probability distribution of time spent in a state, given the destination state. For example, the density of sojourn time in H conditional on a future transition to P is

$$\bar{w}_{\mathrm{hp}}(t) = \frac{w_{\mathrm{hp}}(t)}{W_{\mathrm{hp}}(0)},$$

while the density of sojourn time in H conditional on a future transition to D is

$$\bar{w}_{\mathrm{hd}}(t) = \frac{w_{\mathrm{hd}}(t)}{W_{\mathrm{hd}}(0)}.$$

Similarly, conditional on a transition to P at time t, the density of sojourn time in P conditional on a future transition to C is

$$\bar{w}_{\mathrm{pc}}(s|t) = \frac{w_{\mathrm{pc}}(s|t)}{W_{\mathrm{pc}}(0|t)},$$

while the density of sojourn time in P conditional on a future transition to D is

$$\bar{w}_{\mathrm{pd}}(s|t) = \frac{w_{\mathrm{pd}}(s|t)}{W_{\mathrm{pd}}(0|t)} = \frac{w_{\mathrm{pd}}(s|t)}{1 - W_{\mathrm{pc}}(0|t)}.$$

The distribution $\bar{w}_{\mathrm{pc}}(s|t)$ is usually called the preclinical sojourn time distribution.

6.3.3 Estimating transition densities via deconvolution

Estimation of transition densities is made difficult by the fact that not all the transitions involved are directly observable. In practice, empirical data are available on the number of women making transitions from P to C, from C to D, and from H or P to D, by age, but transitions from H to P are not directly observable. Next we discuss an approach to estimating the distributions w_{hp}, w_{hd}, w_{pc} and w_{pd} in this situation, based on Parmigiani and Skates (2001).

If no screening takes place, $y = t + s$ is the age of the patient at the time of symptoms. The instantaneous probability of making a transition from P to C at age y, or $I_{\mathrm{c}}(y)$, is the incidence of symptomatic disease. This function can be expressed in terms of transition densities via the simple convolution relationship

$$I_{\mathrm{c}}(y) = \int_0^y w_{\mathrm{hp}}(t) w_{\mathrm{pc}}(y - t|t) dt, \qquad (6.1)$$

based on the fact that in order to become incident at y, a woman needs to arrive in P at age t, and leave for C after $y - t$ years, with t being any age prior to y.

Similarly, the probability density of dying from causes other than breast cancer at age y, prior to a breast cancer diagnosis, is the function $I_{\mathrm{d}}(y)$, given by

$$I_{\mathrm{d}}(y) = w_{\mathrm{hd}}(y) + \int_0^y w_{\mathrm{hp}}(t) w_{\mathrm{pd}}(y - t|t) dt, \qquad (6.2)$$

where the two terms correspond to individuals dying while in H and P respectively.

Our approach to the estimation of w_{hp}, w_{hd}, w_{pc} and w_{pd} is based on combining estimates of the following three densities:

1. The density of new clinical cases as a function of age $I_c(y)$, which is directly observable and well documented in cancer registries; in our analysis we will use results of an analysis published by Moolgavkar *et al.* (1979), who developed estimates of the the incidence of breast cancer, accounting for both age and cohort effect.

2. The density of mortality from other causes prior to diagnosis, $I_d(y)$. Overall mortality from causes other than breast cancer is typically available from life tables; to obtain $I_d(y)$ we need to subtract cases previously diagnosed with breast cancer. The distribution of these cases by age is empirically observable from case series of breast cancer patients.

3. The density $\bar{w}_{pc}(u|t)$, of the time in P conditional on moving to C and on having moved to P at time t. This can be estimated based on screening trials data or on screened cohorts. In breast cancer, definitive evidence is still lacking. Our results are based on considering in turn three alternative assumptions, discussed in Section 6.3.4.

It is difficult to gather data on preclinical sojourn time for individuals dying of other causes. However, it is reasonable to assume that individuals in P are subject to the same risks of death from causes other than cancer as the aggregate of individuals in states H and P. This is generally not a restrictive assumption, because dying as the result of having preclinical disease without any symptomatic manifestations is a rather rare occurrence. Formally, if we define

$$K(y) = \int_y^\infty I_d(y')dy',$$

the condition above can be expressed as

$$\bar{w}_{pd}(s|t) = \frac{I_d(t+s)}{K(t)}.$$

Because of the independence assumption just stated, the probability $W_{pc}(0|t)$ of reaching C conditional on entering P at age t can be thought of as the minimum of two independent random variables with distributions $\bar{w}_{pc}$ and $\bar{w}_{pd}$. So we can express $W_{pc}(0|t)$ in terms of known functions, specified in steps 2 and 3 above, as:

$$W_{pc}(0|t) = \int_0^\infty \bar{w}_{pc}(u|t) \int_u^\infty \frac{I_d(s+t)}{K(t)} ds du \qquad (6.3)$$

$$= \frac{1}{K(t)} \int_0^\infty \bar{w}_{pc}(u|t) K(u+t) du.$$

Rewriting expression (6.1) as

$$I_c(y) = \int_0^y w_{\mathrm{hp}}(t) W_{\mathrm{pc}}(0|t) \bar{w}_{\mathrm{pc}}(y - t|t) dt, \qquad (6.4)$$

the only remaining unknown in expression (6.4) above is w_{hp}. Solving this integral equation gives the distribution w_{hp} required by the natural history model. A practical approach is to discretize (6.4) and reduce it to a matrix inversion problem. Because the inversion is typically ill behaved, a singular value decomposition followed by thresholding of low eigenvalues is useful (Parmigiani and Skates, 2001). Once that is done, w_{hp} and w_{pd} can be used to determine w_{hd}.

6.3.4 Alternative estimates of preclinical sojourn time

Estimation of $\bar{w}_{\mathrm{pc}}$ using screening data is a challenging problem, and has drawn much methodological attention (Albert *et al.*, 1978; Day and Walter, 1984; Brookmeyer *et al.*, 1986; Etzioni and Shen, 1997; Straatman *et al.*, 1997). In breast cancer, several studies have addressed this question, but the results are not sufficiently consistent to encourage a meta-analysis, partly because differences arise from distributional assumptions that are not easily validated empirically. Our approach here is based on considering three alternative scenarios, each based on a different study, and then conducting a scenario-based sensitivity analysis. Estimation in each study has not resolved all uncertainties. In practical terms, we can think of $\bar{w}_{\mathrm{pc}}(s|t)$ as a marginal distribution after unknown parameters have been integrated out. These are the three scenarios we consider.

1. *Exponential.* Several authors have used an exponential specification in estimating the sojourn time distribution. Straatman *et al.* (1997) estimated $\bar{w}_{\mathrm{pc}}$ to be exponential with mean 2.3. Walter and Day (1983) and Day and Walter (1984) also used the exponential to analyze data from the Health Insurance Plan (HIP) of Greater New York randomized study of breast cancer screening (Shapiro *et al.*, 1988). They provide a comparison with alternative and more highly parameterized distributions, failing to show an improved fit. While it is difficult to demonstrate its limitations empirically due to the weak and somewhat indirect evidence provided by screening studies, the assumption of exponential duration in state P has limitations in this context, such as the implausible assumption of a mode at zero, and the fast decay of the tail, not adequately accounting for slow-growing tumors. An additional drawback of using HIP data for model selection and estimation of sojourn time is that mammography is now more accurate than it was at the time the trial was conducted. This affects the definition of state P: with current mammography, tumors become detectable earlier so that t tends to be smaller and s larger.

2. *Spratt.* Spratt *et al.* (1986) report estimates of the sojourn time in the preclinical stage by age of the patient at diagnosis, of importance in this

problem. They consider age at diagnosis, while our model requires age at onset of detectable disease. However, they utilize eight broad age groups, so that this difference should not be a concern. The authors postulated a lognormal distribution for the preclinical sojourn time, that is,

$$\bar{w}_{\mathrm{pc}}(s|\mu(t), \sigma(t)) = \frac{1}{\sqrt{2\pi}\sigma(t)s} \exp\left\{-\frac{1}{2\sigma^2(t)}(\log s - \mu(t))^2\right\},$$

where $\mu(t)$ and $\sigma(t)$ vary over the eight age groups considered. The choice of the lognormal is motivated by independent evidence regarding the growth rate of tumors.

In our analysis, we use a slight elaboration of their specification. First, it is convenient to specify a sojourn time distribution indexed by an arbitrary age t, and it seems appropriate to stabilize the location parameter $\mu(t)$ and make it monotone in t. To this end, we modeled $\mu(t)$ as a function of the age t, using

$$\log(m_0 - m(t)) = m_1 + m_2 t.$$

We specified m_0 to approach linearity and constant variance in the regression; an effective choice was $m_0 = 1.4$. This approach yielded least squares estimates of $m_1 = 1.6$ and $m_2 = -0.038$. Second, the variation of $\sigma(t)$ over age intervals failed to show an age trend, and appeared noisy, so we used the eight values to estimate a second-stage prior distribution for a single parameter σ common to all age groups. We chose an inverse gamma density (Gelman *et al.*, 1995, p. 475) with parameters $\alpha = 1.50$ and $\beta = 3.36$, obtained by matching the moments of the reported age-specific variances.

After marginalization over σ, the distribution of s given t is a scale mixture of lognormals, given by:

$$\bar{w}_{\mathrm{pc}}(s|t) = \frac{B(a, 0.5)}{s} \sqrt{\frac{b}{2}} \left[1 + \frac{b}{2}\left(\log s - m_0 + e^{m_1 + m_2 t}\right)\right]^{-\frac{2a+1}{2}},$$

where B is the beta function. The resulting family of sojourn time distributions is shown in Figure 6.3, in the top panel. The effect of age is both to shift the distribution toward the right and to increase its variability. The heavy tail near 0 is the result of integrating out the unknown σ.

3. *Peer.* Peer *et al.* (1993; 1996a) estimated the growth rate of breast cancer using data from women undergoing screening in the Nijmegen trial. They considered three age groups: under 50, 50–70, and over 70. They estimated that the median doubling time of the volume of primary breast cancers was 80, 157, and 188 days in the three age groups. The 95% confidence limits were 44–147 days, 121–204 days, and 120–295 days, respectively. Assuming that the smallest tumor detectable by mammography has a diameter of 5 mm and that the mean diameter at which symptoms develop is 20 mm, and postulating

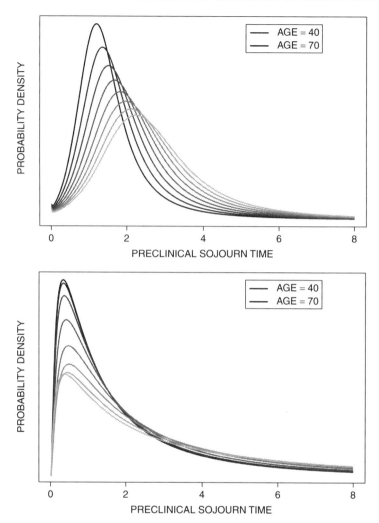

Figure 6.3 Duration of preclinical disease by age under specifications 2 (top) and 3 (bottom). Each curve is an estimate of $\bar{w}_{\mathrm{pc}}(s|t)$ for 10 different choices of age of preclinical onset t, ranging from 40 (darker shade, least dispersed), to 70 (lighter shade, most dispersed). The preclinical sojourn time s is on the horizontal axis.

exponential growth, they use the distribution of growth rates to develop a distribution of time in C. We used their results to approximate the predictive distribution of sojourn time for individual patients. Implicit in their model is an exponential sojourn time distribution. We used this and the confidence limits reported above to estimate the upper 95% quantile of the predictive distribution in the three intervals above. We interpolated the values to obtain a smooth function in the interval from 50 to 70. Then we fit the resulting predictive distribution of sojourn time using a lognormal with median and upper 95% quantile matching Peer *et al.* (1993).

The resulting family of sojourn time distributions is shown in Figure 6.3, in the bottom panel. As age increases, the mode of the distribution remains relatively stable, while the right tail increases. The medians of the resulting curves range from 1.3 (age 40) to 3.1 (age 70). The two curves of Figure 6.3 represent rather different biological mechanisms: in the top curve, most older women have slower-growing tumors; in the bottom curve, only a fraction of the older women have slower-growing tumors, but those who do experience significantly lower growth rates. As these families of curves have potentially different implications for screening, it is interesting to investigate the effects of alternative choices by scenario-based sensitivity analysis, rather than formally combining them.

6.3.5 Age of onset of preclinical disease

We can use each of these estimates of the preclinical sojourn time distribution, and the deconvolution approach of Section 6.3.3, to estimate w_{hp}, w_{hd}, w_{pc}, and w_{pd}. Figure 6.4 illustrates the results. The dotted line is the empirical incidence of clinical disease I_c, the continuous line is the incidence of preclinical disease w_{hp}, the dashed line is the density w_{pd} of transitions from P to D. Scenario 2 predicts a significantly higher number of preclinical cases than the other two, but also has difficulties fitting empirical distribution for older ages, and had to be truncated at 0 after age 90. This suggests that scenario 2 is not perfectly compatible with the empirical incidence I_c.

The area under the w_{pd} curve is the proportion of women in the cohort who have had detectable cancer that never became symptomatic due to competing risks. In terms of the proportion of all detectable breast cancers, this represents 18% (left), 23% (center), and 14% (right), depending on the scenario. The overall number of predicted asymptomatic cases is sensitive to the specification of the sojourn time distribution.

The prevalence of preclinical breast cancer at any given age y is a convolution of the women entering P at any age t prior to y and remaining in P for at least $y - t$. The prevalence for the subset of women who die from other causes after entering state P is derived by conditioning on exiting to D. The ratio of the two prevalences, shown in Figure 6.5, gives insight into the upper age limit of suitability for screening. At age 80 a large proportion (41% in scenario 1, 44% in scenario 2, and 38% in scenario 3) of women will develop preclinical disease, but will die of other causes before becoming symptomatic. The detection of breast cancer in such women will not be of benefit, and could result in unnecessary morbidity, anxiety, and costly medical interventions. An additional proportion of women will die of other causes after developing symptoms. Estimates of the quantity at age 80 show relatively low sensitivity to the choice of sojourn time distribution.

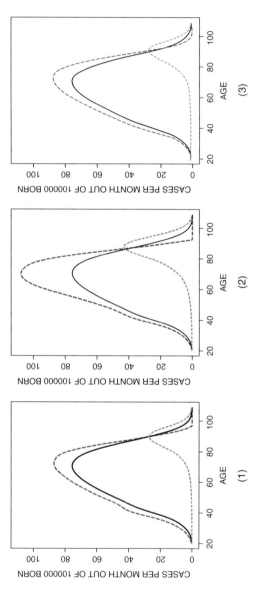

Figure 6.4 Density of clinical and preclinical breast cancer under specifications 1 (exponential), 2 (Spratt), and 3 (Peer). The continuous line is the empirical density of clinical disease I_c and is the same in all panels. The upper dashed line is the density w_{pc} of transitions into the detectable preclinical state P. The lower dashed line is the density w_{pd} of transitions from P to D, for cases where death from other causes occurs after the onset of preclinical disease, but before symptoms.

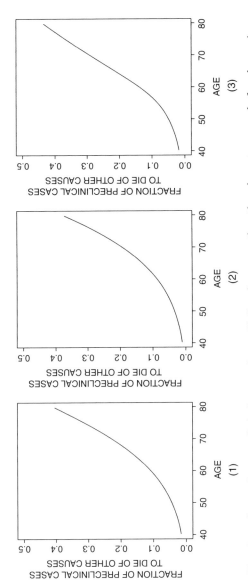

Figure 6.5 Fraction of prevalent preclinical cases that will die of causes other than breast cancer before becoming symptomatic in an unscreened population, by age. The denominator of this fraction is the estimated prevalence of preclinical disease by age. As earlier, we carried out the analysis under specifications 1 (exponential), 2 (Spratt), and 3 (Peer).

6.3.6   Other risk factors

The analysis developed so far refers to a generic patient with no information available about risk factors, aside from age. In practice, more individualized risk prediction and prognosis could affect decision making, especially if based on genetic susceptibility, biomarkers, or prior history of benign breast disease. Some components of the natural history model presented here could be refined by the incorporation of covariates. One example is the distribution w_{hp}, which could be connected to a model of breast cancer risk. Currently, the most widely utilized tool for individualized absolute risk prediction is a model developed by Gail and his colleagues (Gail *et al.*, 1989; Gail and Benichou, 1994; Benichou *et al.*, 1997). This is based on a screened cohort, and predicts $t + s$, but with some adaptation and extrapolation it could be integrated into the natural history model.

In other areas our knowledge of the role of covariates is still too vague to develop reliable generalizations. For example, while it is likely that environmental exposures and reproductive history may affect the duration of the preclinical period, we do not know enough to incorporate these effects into the distribution of $\bar{w}_{\text{pc}}$. As another example, increased mammography has been suggested as a potentially useful strategy for women who are at high risk of breast cancer because of inherited susceptibility. Some of these women have a lifetime probability of developing breast cancer in excess of 50%, and tend to develop cancer earlier than normal women, as we saw in Section 1.3.1. For them, the projected increase in life expectancy from screening could be substantial. However, there is some evidence that increased mammography may be harmful, because of a possible interaction between mutations of the *BRCA1* and *BRCA2* genes and the *RAD51* gene (Scully *et al.*, 1997). Modeling this interaction would require postulating a relationship between mammography and breast cancer risk in genetically susceptible women, which, at the moment, appears to be difficult because of the paucity of direct evidence.

6.4   Modeling the effects of screening

6.4.1   Sensitivity model

Mammography gives both false negative and false positive results. A false negative occurs when a mammogram misses an existing detectable tumor. The sensitivity β is the probability of a positive screening mammogram given a detectable tumor, that is, given that the woman is in state P. Eddy *et al.* (1989) estimate β to be between 0.95 and 0.97. Estimates based on the technology of the late 1970s are lower, around 0.8 (Brookmeyer *et al.*, 1986). Evidence from longitudinal analyses of screened cohorts suggests that the sensitivity may vary with the age of the woman (Peer *et al.*, 1996a). Sensitivity is likely to depend on the size of the tumor at the time of examination. There is also variation in mammography results depending on the radiologist reading the film. Kerlikowske *et al.* (1998) found only a moderate agreement between radiologists in reporting the presence of a finding when cancer was present (kappa = 0.54) and slightly better agreement when cancer was not present (kappa = 0.62).

In our analyses we consider two scenarios. In the first scenario the sensitivity is independent of age and tumor size, but could vary from woman to woman, and is uniformly distributed between β_0 and 1. Random variation reflects both tumor heterogeneity and other subject-specific features that affect the ability of mammography to detect breast cancer, such as breast density. β_0 is taken to be either 0.7, or 0.8, so that the average sensitivity in the two scenarios is either 0.85, or 0.9, respectively.

In the second scenario we consider the case in which β depends on age and preclinical sojourn time. Specifically, we assume that the sensitivity is a smooth increasing function of both age and tumor size, given by

$$\text{logit}(\beta) = -2.2 + 4.4 \, (-0.5) + .11(\tau - 60).$$

Here τ is the age of the patient at the time of the examination, in years, and d, the diameter of the tumor at examination, is measured in centimeters. This is a hypothetical case, loosely based on Peer *et al.* (1996a) and Rosenberg *et al.* (1998). Coefficients are determined to obtain a sensitivity of 0.1 for a diameter of 0.5 cm at age 60; a sensitivity of 0.9 for a diameter of 1.5 cm at age 60; and a sensitivity of 0.64 for a diameter of 1 cm at age 45. This scenario is likely to provide a conservative lower bound to the sensitivity. In our simulation, tumor size at the time of the examination is determined from the current preclinical sojourn time assuming that the tumor has a diameter of 0.5 cm when the woman enters state P, that the diameter at which symptoms develop is 2 cm (Peer *et al.*, 1996a), and that tumor growth is exponential.

In both scenarios we assume, somewhat restrictively, that the occurrences of false negative results in successive mammograms of the same patient are independent. This assumption would be violated if, as plausible, lesions that are hard to detect would remain so from one mammogram to the next.

A false positive occurs when a woman who does not have cancer has an abnormal mammogram. This requires a follow-up procedure, usually a biopsy. False positives are common. Elmore *et al.* (1998) estimate that the probability of at least one false positive result after ten mammograms is 49.1%. Usually, positive mammograms followed by a negative biopsy have generally no immediate consequence and the woman will continue screening at later times. False positives represent a substantial additional cost.

6.4.2 Axillary lymph node involvement model

The axillary lymph node involvement model consists of a conditional probability distribution for the number of axillary lymph nodes that are found to involve metastases. This is a strong indicator of the presence of disease elsewhere in the body as well. The model is specified using empirical cell probabilities by age and tumor size from the SEER registries (National Cancer Institute, 1997).

6.4.3 Survival model

The survival model predicts length of life after diagnosis. We consider four prognostic factors: tumor size at diagnosis; age at diagnosis; the number of surgically found metastases in the axillary lymph nodes; and whether the tumor is estrogen receptor (ER) positive. ER status is assumed to be independent of the other factors. In addition to depending on prognostic factors, survival and QOL depend on treatment. Here we assume that patients are treated according to the guidelines developed by the National Institutes of Health Consensus Conference on Early Breast Cancer (1991). Cases that are unresolved or left to the woman's choice in the Consensus conference are resolved based on the decision analysis of Chapter 5. Women diagnosed before age 50 are treated with Tamoxifen if node-negative and ER positive and with chemotherapy otherwise. Women diagnosed after age 50 are treated with Tamoxifen if node-negative and ER positive, and with chemotherapy followed by Tamoxifen otherwise. The main difference in treatment resulting from an early detection stems from the fact that screen-detected cases present a lower percentage of cases with positive axillary node involvement. Consequently, screen-detected cases will receive chemotherapy less often, with a resulting improvement in QOL. Most other treatment decisions, such as those related to treatment of metastatic breast cancer after recurrence, are unlikely to depend appreciably on screen detection, and are not considered.

As in Chapter 5, survival of patients with positive nodes receiving chemotherapy was estimated using a Cox regression model with age, ER status, primary tumor size, and nodal involvement as prognostic variables, based on the combined CALGB studies 7581, 8082, and 8541 (Wood etal., 1985; Perloff *et al.*, 1996; Wood *et al.*, 1994). This defines the probability distribution $w_{cd}(v|t, s, x)$. The benefits of different adjuvant therapies in terms of hazards of death are also the same as in the ALND decision analysis of Chapter 5.

The distribution w_{cd} depends on screening only via the prognostic variables t, s, x. We also assume that if detection occurs via screening at time τ, the woman will be free of disease-related mortality for the remainder of the (unobserved) sojourn time in P, that is, for at least $t + s - \tau$ more years. This means that a screen-detected subject will not die of cancer earlier than she would have become symptomatic if unscreened. This assumption is not restrictive unless there is a mortality associated with treatment immediately following diagnosis, which is negligible in breast cancer. It can also be relaxed trivially by allowing for negative values of v, although that is not done here.

The reason for this specification is to define the variable v as the additional length of life after $t + s$, so that the interpretation of state C under different detection modalities remains the same.

6.4.4 Costs

We capture the costs associated with screening via the expected number of screening examinations to be performed. Costs should be roughly proportional to the number of examinations. There may be savings in adjuvant therapy and, when screening is successful, patients saved from cancer by early detection may incur further medical expenses later on in their lives. These additional cost implications of screening are not factored into our analysis. To give a sense of the magnitudes involved, a mammogram can cost from US$50 to $200. The biopsy that follows it in the case of a positive outcome costs around $1000. Approximately one in five mammograms in the USA are false positives. False positives create a brief but significant distress, so that changing the expected number of examinations has QOL implications. For further discussion of costs, see Hillner and Smith (1991) and Lindfors and Rosenquist (1995).

As we discussed in Chapter 2 from the perspective of societal decision making, trade-offs between costs and their net health impacts can be analyzed via a cost-effectiveness analysis, comparing the incremental costs of a health intervention and the incremental health benefits that result from it. Our analysis will focus on representing strategies in the trade-off plane, a representation can be used to perform an analysis similar to cost-effectiveness with respect to any reference strategy. Cost-effectiveness of breast cancer screening has been frequently addressed in the literature (Lindfors and Rosenquist, 1995; Brown, 1992; Mushlin and Fintor, 1992; van Ineveld etal., 1993; de Koning *et al.*, 1991; Brown and Fintor, 1993; Elixhauser, 1991; Clark, 1992; Parmigiani and Kamlet, 1993; Eddy *et al.*, 1989; Carter *et al.*, 1993; Saltzmann *et al.*, 1997).

6.5 Comparing screening schedules

6.5.1 Screening strategies

A schedule of screening examination times is indicated by the sequence $\tau = \{\tau_i\}_{i=1,2,...}$, where τ_i is the age of the patient at the ith examination. In this section we will focus on periodic schedules, which are defined by three quantities: the age a_0 at which examinations begin, the period a_1 between successive examinations, and the age a_2 beyond which no further examinations take place. So the first examination takes place at age $\tau_1 = a_0$ the second at age $\tau_2 = a_0 + a_1$, and the ith at age $\tau_i = a_0 + (i-1)a_1$, as long as $\tau_i < a_2$. The generic examination time will sometimes be denoted by τ with no subscript, when this generates no confusion. We assume that all examinations are carried

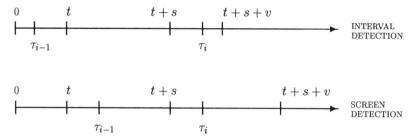

Figure 6.6 Screen detection versus interval detection.

out as scheduled. This is appropriate for analyzing individual decision making, but less so for analyzing the impact of community-based intervention, in which women may skip examinations.

To illustrate the difference between detection by symptoms (also called interval detection, because women present with the illness in the interval between examinations) and by screening (screen detection) consider Figure 6.6. In the top line (interval detection) breast cancer progresses through the preclinical stage within the interval between two successive mammograms, and is detected by symptoms at time $t+s$. The patient survives an additional v years. In the bottom line (screen detection) a mammogram is scheduled during the preclinical sojourn time and is successful at detecting cancer. Diagnosis occurs at time τ_{i-1}. The patient survives is an additional $v + t + s - \tau_{i-1}$ years. In this example screen detection is associated with a longer survival. While this is the expected direction of the inequality, v is random under both scenarios, and the opposite could occur.

A common concern in the evaluation of screening programs is lead time bias (Zelen, 1976). This occurs, for example, when the evaluation criterion is survival since diagnosis. In that case screening has an unfair advantage over not screening, as it would appear to be advantageous even for women who would die at age (say) 60, irrespective of mode of detection. If we consider the total length of life, quality-adjusted or not, issues of lead time bias are avoided altogether.

Individual decisions about screening involve highly subjective trade-offs. A formal decision analysis could be carried out by eliciting individual values, converting them into QOL measures, and then providing women with the results. This, however, is unlikely to work well because of the complexity of the utilities required, and may also create the impression of moving the decision center from the woman to a model that may not represent her values faithfully enough. Alternatively, women could be shown results of an analysis based on QOL assessments elicited from other women who experienced the relevant health states. This may imply cross-cultural or cross-generational comparisons, which are potentially problematic. However, women sometimes inquire explicitly about the QOL experienced by others.

An open issue concerns the appropriate way of presenting policy makers and individual women with quantitative information that can effectively help them address these trade-offs. Here we present two approaches: a balance sheet approach in Section 6.5.2 and a trade-off analysis in Section 6.5.3.

6.5.2   Balance sheet

A balance sheet is a summary of the expected benefits of and harm caused by an intervention. Its goal is to inform decision makers, and enable them to weigh benefits and harm according to their individual values. This approach has been advocated as the most appropriate way of using decision models in medicine, and has been applied successfully to stroke, breast cancer, and elsewhere (Matchar and Samsa, 1999; Barratt et al., 1999).

Table 6.1 is an example of a balance sheet for breast cancer screening, based on the model of this chapter. We consider two strategies: annual screening between ages 40 and 75; and annual screening between ages 50 and 75. Figures are obtained using 12 000 Monte Carlo replicates simulated from scenario 3 for preclinical sojourn time and using the logistic sensitivity model. This is a large number of simulations but does not completely remove Monte Carlo error. Figures have been rounded for ease of communication. All figures are relative to no screening. For example, 0.456 is the difference between the life expectancy in years in the screened and unscreened cohort.

Differences between the two columns can inform decision makers about screening before age 50. The additional number of examinations per woman is about 9, as expected. Elmore et al. (1998) report on a retrospective cohort study of breast cancer screening and diagnostic evaluations among 2400 women who were 40–69 years old at study entry. False positive results occurred in 6.5% of the mammograms. Using this estimate, screening in the forties leads to 0.6 additional false positives per woman. Of interest is also the

Table 6.1 Balance sheet for two alternative screening strategies: annual screening starting at 40 and at 50 years of age. In both cases screening stops at age 75. All figures are increments compared to no screening for a cohort of 10 000 women.

	Starting age 40	Starting age 50
Additional number of examinations per woman	34	25
Additional number of false positives per woman	2.2	1.6
Additional years of life per woman	0.456	0.423
Additional years as cancer survivor per woman	2.23	2.16
Additional women detected with no nodal involvement	279	278
Women treated unnecessarily	30	30

probability of experiencing a false positive. Elmore *et al.* (1998) estimate that the cumulative risk of a false positive result is 49.1% (95 percent confidence interval 40.3–64.1%) after 10 mammograms.

The difference in expected years of life amounts to about 12 days, with a Monte Carlo standard deviation of about 2 days. This comes at a price of about 25 additional days spent as a cancer survivor, which for many women involves a lower QOL. A very small fraction of women, 1 in 10 000 in this simulation, will avoid having lymph node involvement. This figure is small as a result of the model assuming that younger women have faster growth rates and lower mammographic sensitivity. This makes it harder to detect tumors early enough to avert nodal involvement. Later screening is far more effective at this specific goal, averting 278 cases compared to no screening.

One of the negative aspects of screening is overdiagnosis. Screening will lead to the detection and treatment of cases that would otherwise never have progressed to the symptomatic stage. This is quantified in our balance sheet by the 'women treated unnecessarily' category. Thirty women out of 10 000 are projected to be overdiagnosed using our model. This figure does not vary with starting age, as overdiagnosis affects primarily older women, who are at higher risk of competing causes of mortality, and also are likely to have slower growing tumors.

6.5.3 Trade-off analysis

The evaluation of a large number of alternative strategies using balance sheets is unwieldy, and graphical representation can be helpful. In this section we illustrate a trade-off analysis, based on summarizing the outcomes of the model in terms of two conflicting objectives of interest, and representing each strategy as a point in the space of these objectives. Rather than combining the two outcomes into a single utility, or using them in a cost-effectiveness framework, a trade-off analysis presents them graphically in a way that may help understand or resolve the conflict. There are several possibilities in the case of screening. In our analysis the objectives of screening are captured by two conflicting outcomes: the number of mammograms carried out, and the quality-adjusted length of life. The resulting trade-off analysis is useful in weighing the combined QALY effects against screening costs, risk of false positives and so on. When the objective space is cost versus effectiveness, a trade-off analysis can be straightforwardly converted into a cost-effectiveness analysis for any pair of strategies in the space.

The QALYs measure considers and weighs each year of life following diagnosis, using the quality of life adjustments of the 'moderate' scenario of Table 5.2. This objective measure resolves internally the trade-off between a longer life, but with an increased portion spent as a cancer survivor, and potential overdiagnosis. Because treatment occurs earlier under screen detection, and may result in permanent loss of QOL, it is possible that such

Table 6.2 Summary of results from eight randomized trials of mammography screening, and estimated mortality reductions.

Trial	Screening group		Control group		Mortality reduction(%)	
	BC* deaths	Life Years	BC* deaths	Life Years	Obs.	Est.
HIP	49	248 000	65	253 000	23	22
Kopparberg	23	144 000	18	75 000	33	22
Östergötland	27	143 000	27	147 000	−3	8
Malmö	57	166 000	78	144 000	38	28
Edinburgh	46	146 000	52	135 000	18	18
Gothenburg	18	138 000	39	168 000	44	29
Stockholm	24	174 000	12	88 000	−1	5
Canada	82	283 000	72	283 000	−14	0
Total	326	1 442 000	363	1 293 000	18	18

*Breast cancer

loss may offset any gains in overall survival, and that increased screening may lead to lower expected QALYs. To study these using a trade-off analysis we could also consider a different objective, say length of life versus lead time.

Figure 6.7 gives an example of trade-off plots. We evaluated the 12 strategies resulting from each combination of $a_0 \in \{35, 40, 45, 50\}$ and $a_1 \in \{0.5, 1, 2\}$. We fixed a_2 at 75 years. We replicated the analysis under sojourn time

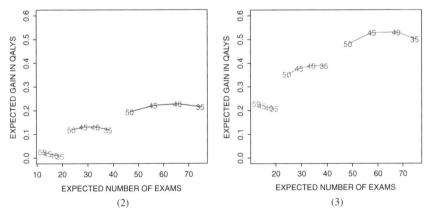

Figure 6.7 Trade-off analysis for the logistic sensitivity model, under sojourn time scenarios 2 (Spratt) and 3 (Peer). Points are plotted with the value of a_0. Strategies with common interval a_1 are connected by lines ($a_1 = 2$ is the leftmost; $a_1 = 0.5$ is the rightmost).

scenarios 2 (Spratt) and 3 (Peer) as a sensitivity analysis. Results are based on Monte Carlo sample sizes of 8000 women. Some smoothing was carried out to correct for the small remaining Monte Carlo variation, as discussed in Section 3.5.1. Each strategy is represented by a point in Figure 6.7. To easily identify the role of the starting age and screening intervals, lines connect all the strategies with the same screening interval, and points are plotted with the relevant starting ages. The opposite labeling scheme is also available, but leads to a cluttered graph in this case.

Strategies that are to the south-east of other strategies are dominated, because they have a lower QALY gain with a higher expected number of examinations. For example, screening every 6 months starting at age 35 leads to more examinations and lower expected QALYs than screening every 6 months starting at age 40. If we focus on undominated strategies in Figure 6.7, we obtain a so-called efficient frontier for the given set of strategies. For a fixed value of the expected number of examinations, the frontier gives the highest QALY that can be obtained within the class of policies considered. Likewise, for a fixed QALY value on the horizontal axis, the frontier gives the lowest number of examinations at which that can be reached. For example, annual screening beginning at 35 years of age is suboptimal by almost 10 examinations, that is, its horizontal distance from the efficient frontier. The richer the set of strategies, the better the efficient frontier. A smooth frontier could be evaluated by varying a_1 and a_2 continuously, and thus generating a family of curves like those of Figure 6.7. The upper left-hand envelope of these curves gives the efficient frontier.

Determining an optimal strategy depends on personal attitudes toward exchanging an expected year of quality-adjusted life with a mammogram. Each exchange rate defines a family of parallel indifference lines in the objectives space. If a_0 and a_1 were allowed to vary continuously, the optimal strategy would be such that one of these lines is tangent to the upper right convex hull at the point corresponding to the strategy. In practice, this exchange rate is difficult to elicit in individual decision making, and trade-off analysis often provide more reliable support to the decision process.

In both scenarios of Figure 6.7 there is a strong effect of the screening interval on both QALYs and number of examinations. The marginal return of decreasing the interval from 1 year to 6 months is only slightly smaller than that of decreasing the interval from 2 to 1 years. The effect of the starting age is far less pronounced, and for shorter intervals decreasing the starting age below 40 is counterproductive, and leads to dominated points. Overall screening is more effective in scenario 3 than in scenario 2, emphasizing the sensitivity of the overall effectiveness to the choice of sojourn time distribution. The gain in QALYs from starting at 40 versus 50 years of age is also affected by the scenario. Screening women between 45 and 50 appears more cost-effective than screening women younger than 45. Mammography every 2 years appears to be too infrequent to yield a benefit in women younger than 40. This hypothesis

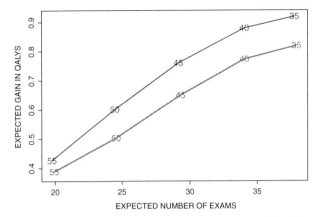

Figure 6.8   Trade-off analysis in the random sensitivity model. The axes are the expected number of examinations and expected gain in quality resulting from each strategy. Each point in the graph represents a screening policy. Points are plotted with value of a_0. All strategies have an interval a_1 of one year. Each line corresponds to a different lower bound β_0 in the distribution of sensitivity: $\beta_0 = 0.8$ is the at the top line and $\beta_0 = 0.7$ is the line at the bottom.

is at present unconfirmed by randomized trial evidence, but is consistent with other analyses (Peer *et al.*, 1996b).

Figure 6.8 shows a similar trade-off analysis to Figure 6.7 for the random sensitivity model. The set of strategies differs: we consider annual screening with starting ages $a_0 \in \{35, 40, 45, 50, 55\}$. As earlier, we fixed the limit age a_2 to 75 years. QALY gains from screening are much increased compared to the logistic sensitivity model, because this model assumes a much greater ability of mammogram to identify small tumors. The advantages of screening in the forties also appear to be much greater, for the same reason. By way of a sensitivity analysis we considered two possible values for the lower bound of the distribution of sensitivity: $\beta_0 = 0.8$ and $\beta_0 = 0.7$. This gives a sense of the likely impact of improvement in mammographic sensitivity.

6.6   Model critique

6.6.1   Limitations of the model

In this section we briefly consider a checklist of points that can be helpful in assessing the validity of a model of this complexity.

A first issue is the quality of the input data that enter the model. Are these data of adequate standard? Is the evidence base sufficiently comprehensive? Is each source used for its best purpose? In the natural history model the critical input is the preclinical sojourn time distribution. We consider three scenarios that cover a relatively wide set of possibilities. Given the difficulty of reliably estimating the preclinical sojourn time distribution based on existing data, these scenarios are likely to be adequately representative of the plausible alternatives.

Both versions of the sensitivity model are somewhat tentative. The assumption that the sensitivity is independent of tumor size and age is likely to be too simplistic. The alternative assumption of a logistic relationship is more plausible, but joint estimation of the effects of age and tumor size has been insufficiently studied, and the coefficients used here are hypothetical. Several studies address sensitivity, either in the aggregate or by age. Mushlin *et al.* (1998) review these studies and perform a meta-analysis. A follow-up combined analysis by age and tumor size would be challenging but could contribute substantially to improving the model presented here.

The axillary lymph node involvement model is based on registry data, generally the most appropriate source for this kind of analysis. Finally, the survival model incorporates evidence from the comprehensive Oxford overviews of randomized clinical trials, and from a large and relevant set of patients from the CALGB studies. A possible improvement would be to use data with similar detail regarding treatment and prognostic factors from a screening trial, if those were available.

A second issue is the plausibility of assumptions. Key assumptions are those about tumor growth rates and their variability, the relation between tumor size and prognosis, and the relationship between detectability and prognosis. We assume that, for a fixed age, small tumors are less likely to have developed metastases, and can be treated more effectively. In particular, the expected gain in survival, based on the prognostic data we use, appears to be larger than the expected lead time in detection. However, there is a possibility that screening will detect a selected subset of these smaller tumor, ones that metastasize earlier, in which case our model-based estimates would overestimate efficacy. There is some indication that this may occur. For example, Vitak *et al.* (1999) investigate whether different mammographic categories of interval cancer, classified according to findings at the latest screening, are associated with different distributions of prognostic factors. The overlooked or misinterpreted tumors showed significantly higher proportions with low S-phase fraction, an indicator of growth rate, and ER positive tumors, which can be treated with hormonal therapy. However, analyses of survival rates disclosed no clear differences between the two radiological categories. Also screening may be harmful, because of irradiation (Feig and Hendrick, 1997), a fact which is not explicitly considered in this model.

A third issue is whether the model is sufficiently detailed to be helpful in informing individual decision making. Results obtained based on the natural history of a 'woman of unknown risk factors', as considered here, may not be informative for a woman who has evidence of being at high risk. We discussed some of the possibilities in Section 6.3.6. Genetic susceptibility, either in the form of a strong family history or a known genetic alteration, is a strong risk factor, and for a genetically susceptible woman, calculations based on the present model may not be very helpful. However, genetic susceptibility affects a relatively small proportion of the population. Other risk factors are not as strong, so that the model's conclusions are likely to be helpful to a large segment of the population.

In addition, two important aspects are still missing at the time of writing. The first is external validation of the projected results. The only indirect evidence is provided by the general agreement of the logistic sensitivity model with clinical trial results. The second is an analysis of uncertainty. The pronounced differences revealed by varying the sensitivity model suggest that parameter uncertainty should be investigated further. From a technical standpoint, a further limitation of all these calculations is that they are 'plug-in' estimates in the sense of Section 2.4.5, with no explicit accounting of parameter uncertainty.

6.6.2 Clinical trials and screening women in their forties

Several randomized clinical trials have studied mammography screening. The National Breast Screening Study (NBSS) I trial (Miller et al., 1997), conducted in Canada was designed explicitly to assess the mortality benefits of screening mammography among women in their forties. There are also additional trials that can provide additional evidence on this issue via post hoc subgroup analyses. Discussion typically focuses on four trials conducted in Sweden (Tabár et al., 1997; Frisell and Lidbrink, 1997; Bjurstam et al., 1997; Andersson and Janzon, 1997), one conducted in Scotland (Alexander, 1997), and one, the Health Insurance Plan (HIP) of New York trial (Shapiro, 1997), conducted in the United States. In a recent review of the evidence from these trials, the Canadian Task Force on Preventive Health Care (2001) observed: 'Although several of the trials reviewed constitute level I evidence (RCT), at present their conflicting results, methodologic differences and, most important, uncertainty about the risk:benefit ratio of screening precludes the assignment of a "good" or "fair" rating to recommendations drawn from them'.

Trials differed in a number of relevant characteristics. The Canadian trial included volunteers, while the other trials were population-based. The Malmö and Edinburgh trials did not include women aged 40–45. Follow-up varies from 10 to 18 years. Randomization was sometimes carried out by assigning women to screening one at a time, and sometimes by assigning to screening clusters of women, say all women who visited a particular clinic or practice. In the Gothenburg study, there was a change in the randomization procedure during the trial; 82% of the women were randomized individually. Reported results also differ depending on the type of analysis (Berry, 1998). Follow-up analysis considers deaths from breast cancer detected at any time after study entry. Evaluation analysis does not consider deaths from breast cancer detected after the study screening period. Primary data are not always available to create valid common scales.

Table 6.2 summarizes mortality results from the eight trials (Hendrick *et al.*, 1997). For each one, the table shows the numbers of deaths attributed to breast cancer and the total numbers of life years by treatment group. Breast cancer mortality rates are measured as deaths per 100 000 life years. Breast cancer mortality reduction in the 'observed' column is $100(1 - r)\%$, where r is the relative risk, that is, the ratio of the mortality rate in the screening group to the mortality rate in the control group. So a negative mortality reduction occurs when the mortality rate in the control group is smaller than in the screening group. The combined relative risk reduction in the 'observed' column is calculated using the Mantel–Haenszel statistic. Mortality rates in the control arms tend to be higher than screening mortality rates and are more variable.

There has been great interest in developing meta-analyses of these trials. Irwig *et al.* (1997) review seven meta-analyses carried out prior to 1997. Estimated relative risk varies from 0.84 to 1.08, with associated confidence intervals 0.69–1.02 and 0.85–1.39, respectively. All confidence intervals include 1. More recently, Berry (1998) performed a meta-analysis using a Bayesian hierarchical approach. The posterior mean of the corresponding study-specific parameters are reported in Table 6.2 as estimated mortality reductions. Berry also estimated the additional life expectancy for screening women in their forties. The screening group has an estimated advantage over the control group of 5.3 days. Similarly, Saltzmann *et al.* (1997) estimate an added life expectancy of 6.5 days.

Our decision model also quantified the same benefits. The magnitude of the benefit depends crucially on the sensitivity model assumed. If sensitivity is independent of age and tumor size, the model predicts a much greater benefit than the trials. If the model incorporates a dependence between the sensitivity and age and tumor size, effects are closer. Scenario 2 gives a difference of 8 days, and scenario 3 a difference of 12 days; 8 and 6.5 days are likely to be within parameter uncertainty. There could be several reasons for this remaining discrepancy between trials and models. On the one hand, model-based comparisons are free of some of the problems that make interpretation of results from randomized trials difficult, such as compliance (women in the control group had the option of seeking mammography screening outside the trial, and did); random starting times (in some trials, the treatment arm started screening women at any age during their forties, possibly diluting the benefit); lack of adherence to schedules; and obsolescence of technology. On the other hand, decision models are based on a constellation of assumptions, some of which are not fully validated, as we argued in Section 6.6.1.

6.6.3 Complementarity of trials and decision models

The comparison just outlined raises a broader methodological question: what are the roles of randomized trials and comprehensive decision models in understanding the efficacy of screening? Neither one alone seems sufficient to address the problems that individuals and policy makers face. There are good reasons for carrying out both, and for integrating the results of one approach into the other. Some points for discussion follow.

It is difficult, and almost never done, to measure individual QOL, and health outcomes other than death or cancer recurrence, in the context of a large prevention trial. Yet individual values need to play a key role in weighing the benefits and risks of screening. Decision models can be used to bring the individual values to bear, by formal utility-based decision analyses, by balance sheets and/or by trade-off plots that enable informed decision making. They can also support, where possible, more formal utility considerations.

Today's mammographic techniques are substantially different from those used in the 1970s, when most of the randomized trials started. For example, the dosage of radiation is smaller and less likely to be harmful, and the quality of the film has improved. As a result, the ability of trials such as the HIP (Shapiro, 1997) to shed light on the important trade-offs could be lessened today. This is a common fate of trials that assess the benefit of prevention technology over a long follow-up. Similar considerations apply prospectively. What is the anticipated benefit of a mammographic procedure that can detect tumors that are half the size of currently detectable tumors? What resources should be devoted to research in this direction? These are questions that need to be addressed quantitatively and rigorously. Decision models offer a framework for this quantification.

In all randomized trials, women were assigned to either a screening program or a control program not involving planned mammography. Mammography is widely available and some patients in the control arms of the trials sought them. Also, in some studies, those assigned to screening were not screened. Trials address the impact of implementing a screening program in a community. They enable solid causal inference, but only in the relatively limited sense of comparing the two interventions as implemented. This is another reason why the results are not as directly applicable to individual decision making.

One of the advantages of decision models is that they can reflect different individual or societal values, and adapt to technological changes and specific implementations of strategies. For this reason they can potentially support decision making more directly and effectively. Modeling relies on both data and assumptions. When assumptions are drawn from solid clinical judgment or earlier studies they constitute an asset to the model. Unfortunately, many models require at least a measure of assumptions that are difficult to validate empirically. By contrast, trial data are meant to be immediately interpretable without resort to further assumptions or models. Unfortunately, in breast cancer screening that has not proven possible, in view of both the limitations of the trial design and implementation, and the gap that exists, even in the best cases, between the trial results and individuals' decisions.

The costs of randomized trials of prevention are prohibitive and divert resources from the development of new diagnostic and therapeutic technologies. A systematic and carefully controlled use of models would enable good inferences from smaller studies and also provide vital guidance in the design of expensive future studies (Flehinger and Kimmel, 1994; Peer *et al.*, 1995; Hu and Zelen, 1997). Simulation of cohorts of patients using the approach of Section 3.3 can be used in designing all aspects of a trial.

Modeling poses substantial challenges of implementation, as we have seen throughout this chapter, but perhaps the most important obstacles to their effective use in medical decision making are in the validation, documentation, and communication of results. Critical assumptions often need to be represented within a highly technical framework, and can be complex to grasp. This makes it difficult for agents who are not technically trained to evaluate complex decision models and rely on them with confidence.

In many cases there is still a wide gap between the quantitative information provided by a decision model and the determinants of individual health behavior, or of policy decisions. This is true for formal utility-based models, but also for models that present evidence tables in a way that does not make any attempt at resolving the conflict for the decision maker. The risk communication literature (Vernon, 1999) provides insight into the issues involved in the translation of quantitative information into behavior modification.

There is also skepticism of decision modeling in the medical community. This is at least in part justified by the prevalence of applications that are of low standard in terms of input, execution, or documentation. With regard to this, Ramsey *et al.* (2000) point out:

> The apparent ad hoc nature of models belies the methodologic rigor that is applied to create the best models in cancer prevention and care. Models have progressed from simple decision trees to extremely complex microsimulation analyses, yet all are built using a logical process based on objective evaluation of the path between intervention and outcome. The best modelers take great care to justify both the structure and content of the model and then test their assumptions using a comprehensive process of sensitivity analysis and model validation.

Progress in the use and understanding of comprehensive models requires that models should be thoroughly documented and made available to decision makers in a sufficient level of detail that the results may be reproduced by others. This would enable such factors as costs, patient profiles, decision objectives, epidemiological and clinical trial evidence to be altered or updated as new information accrues. Good models have the potential of becoming highly structured ways of organizing and keeping up to date the information about a particular disease.

In summary, clinical trials have a solid foundation and, when conclusive, a compelling appeal, but in many cases are not sufficient to inform decision makers facing the complexities of practical medical problems. Trials are useful for estimating intervention effects but are less apt at producing the predictions that are more directly relevant for informed decision making. Models provide a useful, and sometimes indispensable, complement. In an ideal situation, trials provide core evidence to enable reliable decision modeling, and models place it in context and translate it into decision-relevant terms.

6.7 Optimizing screening schedules

The final section of this chapter is a digression into the more technical topic of finding screening strategies in which the interval between examinations can vary with age. The pioneering work of Barlow et al. (1963) and Lincoln and Weiss (1964), started a tradition of decision-theoretic approaches for finding the optimal sequence of checking times. Several later development have addressed optimal examination times for medical checkups or screening examinations, (Lincoln and Weiss, 1964; Weiss and Lincoln, 1966; Shahani and Crease, 1977; Eddy, 1983; Tsodikov and Yakovlev, 1991; Parmigiani, 1993; 1997; Zelen, 1993; Tsodikov et al., 1995), and have applied optimality ideas to specific diseases (Kirch and Klein, 1974; Lashner et al., 1988; Lee and Zelen, 1998; Parmigiani, 1998). In this section we briefly review two approaches, a recursive approach for the determination of exact solutions, and a variational approach for developing approximate solutions.

6.7.1 Recursive approach

This section considers the general statistical design problems of choosing examination ages in screening for chronic disease. The focus is on capturing the trade-off between the costs of examination and the losses incurred as a result of a delay in detection. This problem can be posed in a decision-theoretic way. We discuss general ideas and the technical details of a simplified formulation based on work by Sengupta (1980). Consider the model of Figure 6.2 and remove transitions from H and P to D. The onset of occult disease occurs at time t and symptoms occur at time $t + s$, where t and s are, say, independent random variables, with known distributions w_{hp} and w_{pc}.

An examination schedule is a sequence $\boldsymbol{\tau} = \{\tau_i\}_{i=1,2,\ldots}$, where τ_i is the time of the ith examination, with $\tau_i > \tau_{i-1}$ and $\tau_0 = 0$. $n = \sup\{i : \tau_i < \infty\}$ is the number of planned examinations, and may be finite or infinite. Screening examinations are administered to asymptomatic individuals and terminate as soon the disease is detected in P, or the individual reaches C and seeks medical advice, or the individual reaches age τ_n. For example, if time is measured in years, the schedule $\boldsymbol{\tau} = \{50, 60, \infty\}$ indicates that the patient – if alive and asymptomatic – is examined at age 50. If the exam is a true positive screening terminates. Otherwise, the patient – if still alive and asymptomatic – is examined again at age 60. Then screening terminates regardless of the patient's state. For a given patient and schedule, the number of examinations actually performed is random, and depends on t and s; it is always smaller than n, and is a proper random variable when n is infinity.

Choosing τ addresses at once the problems of timing examinations, of which segment of the population to screen, and of when to stop screening. In particular, if n is 0, individuals with the postulated transition probabilities are not to be screened; if n is positive and finite, screening times are $\tau_1, \ldots, \tau_n$, and if the individual is alive after τ_n, screening stops; if n is infinite, the individual is screened for the remainder of his or her life.

Schedules are chosen based on expected utility analysis. Here, by way of illustration, we assume that each examination costs a fixed amount k, on the utility scale, independent of the age of the patient, the stage of the disease, and the number of previous examinations. We also assume that benefits of early detection can be represented by the lead time in detection. Each year of delay in detection costs an amount c, again on the utility scale.

For a given schedule τ, define $\mathcal{N}(\tau)$ to be the expected number of examinations and $\mathcal{L}(\tau)$ to be the expected delay in detection. Expectations are taken with respect to the joint distribution of t and s. The optimal schedule is chosen to maximize the total expected utility:

$$\mathcal{U}(\tau) = -k\,\mathcal{N}(\tau) - c\,\mathcal{L}(\tau). \tag{6.5}$$

From (6.5) we see that the optimal τ depends on k and c only through k/c. This simplifies matters, although specifying k/c can be hard. However, techniques to optimize (6.5) are useful even if k/c is not specified. Consider minimizing $\mathcal{L}(\tau)$ for a fixed expected number of examinations $\mathcal{N}_0$, for example to find the best way to allocate a fixed expected budget. The Lagrangian is

$$\ell\left[\mathcal{N}(\tau) - \mathcal{N}_0\right] + \mathcal{L}(\tau),$$

which is proportional to (6.5). If the objective function is sufficiently regular, and extrema can be found by solving first-order conditions obtained by setting first derivatives to zero, then the first-order conditions for the unconstrained optimization of $\mathcal{U}$ are the same as those for the constrained optimization of $\mathcal{L}$, with k/c replaced by ℓ. Choosing $\mathcal{N}_0$ in the constrained optimization of $\mathcal{L}$ is alternative to eliciting k/c in the unconstrained optimization of $\mathcal{U}$.

Graphing $\mathcal{N}_0$ against $\mathcal{L}(\tau_0)$, where τ_0 is the solution of the constrained optimization, gives the efficiency frontier of the trade-off plot of Figure 6.7. Graphing the efficiency frontier does not require the ratio k/c to be specified, and provides decision makers with an informative view of both average and marginal costs of screening examinations, and of whether any given schedule is inefficient. In constructing such a plot, the axes can also be chosen differently.

For example, if both $\mathcal{N}$ and $\mathcal{L}$ include financial as well as QOL measurements, it can be more natural to aggregate by these units and plot dollars versus QALYs. See Parmigiani and Kamlet (1993) for an efficiency frontier for screening for breast cancer.

In the simple formulation laid out so far, the expected utility of not screening is $-cE(s)$. If this is greater than the expected utility associated with the best examination schedule, then the optimal action is not to screen. Assuming that at least one examination is scheduled, and that the sensitivity β is 1, the expected utility of schedule $\boldsymbol{\tau}$ is:

$$
\mathcal{U}(\boldsymbol{\tau}) = -\sum_{i=0}^{\infty} \int_{\tau_i}^{\tau_{i+1}} \left[\int_0^{\tau_{i+1}-t} (ki + cu) w_{\mathrm{pc}}(u) du \right.
$$

$$
\left. + \int_{\tau_{i+1}-t}^{\infty} [(i+1)k + c(\tau_{i+1} - t)] w_{\mathrm{pc}}(u) du \right] w_{\mathrm{hp}}(t) dt. \quad (6.6)
$$

Of the two terms in the square brackets, the first corresponds to interval detection, the second to screen detection.

By computing partial derivatives of $\mathcal{U}$ with respect to each of the τ_i, and setting them to zero, we can obtain sufficient conditions for the optimality of schedule $\boldsymbol{\tau}$. These have the following form:

$$
W_{\mathrm{pc}}(\tau_{i+1} - \tau_i) + \frac{k}{c} \int_0^{\tau_{i+1}-\tau_i} W_{\mathrm{pc}}(s) ds
$$

$$
= \int_0^{\tau_i - \tau_{i-1}} \frac{w_{\mathrm{hp}}(\tau_i - s)}{w_{\mathrm{hp}}(\tau_i)} \left[W_{\mathrm{pc}}(s) - \frac{k}{c} w_{\mathrm{pc}}(s) \right] ds, \quad i \geq 1, \quad (6.7)
$$

where $W_{\mathrm{pc}}(s)$ is a tail probability as defined in Section 6.3.2. Assuming that the hazard rate of the preclinical sojourn time distribution, $w_{\mathrm{pc}}(s)/W_{\mathrm{pc}}(s)$, is bounded by c/k, each equation in (6.7) can be used to solve for the optimal τ_{i+1} as a function of τ_i and τ_{i-1}. So given any initial examination time τ_1, equations (6.7) can be used to recursively determine the optimal continuation schedule. This reduces our infinite-dimensional optimization to a one-dimensional problem: determining the best τ_1.

A condition that has played a critical role in the analysis of optimal schedules is log concavity of $w_{\mathrm{hp}}(t)$, that is, the condition that the logarithm of $w_{\mathrm{hp}}(t)$ is a concave function. All log concave densities have increasing failure rate, which makes them suitable for modeling chronic diseases; examples include truncated normal, exponential, Weibull (shape parameter greater than one), gamma (shape parameter greater than one) and Gumbel (Barlow and Proschan, 1965). Sengupta (1980) considered a log-concave density for t, and derived an elegant result concerning the stopping rule. Let

$$
\varphi(\delta, t) = \int_0^\delta \frac{w_{\mathrm{hp}}(t - s)}{w_{\mathrm{hp}}(t)} \left[W_{\mathrm{pc}}(s) - \frac{k}{c} w_{\mathrm{pc}}(s) \right] ds. \quad (6.8)
$$

Then if $\lim_{t \to \infty} \varphi(t, t) \le cE(s)$ it is optimal not to screen. Otherwise, the optimal inspection schedule consists of an infinite number of examinations. He also showed that under the same conditions, intervals between examinations should be increasing, consistently with the increasing failure rate of t, and the optimal schedule is unique and can be determined by a practical binary search algorithm that does not required evaluation of $\mathcal{U}$, but only repeated solution of the equations in (6.7).

Tsodikov and Yakovlev (1991) and Parmigiani (1993) consider various extensions of these results, to account for competing causes of death, dependencies among t and s, and more general objective functions. Some results extend also to the case of $\beta < 1$, which in general, however, loses the nice recursive structure provided by (6.7).

6.7.2 Variational approach

In this section we present an alternative approach for choosing examination times based on approximating the decision space, which is a space of sequences, using a space of functions (Keller 1974). Methods for finding optima in function spaces, or variational methods, have a long tradition in statistical design (Rustagi, 1976) and can also be applied successfully in this context. This discussion is based on Parmigiani (1997).

One of the advantages of variational approaches is speed of computation. Exact recursive solutions for examination schedules are computer-intensive when the screening test has sensitivity less than unity, as they require the solution of a sequence of nonlinear systems of equations, evaluation of which requires numerical integration. For instance, in breast cancer screening, systems with 10–30 unknowns are common, and can typically lead to about 100 000 numerical integration steps. Using a variational approach, a good approximate solution to the same problem is available by simple numerical inversion of one known function. In nontrivial special cases, the inverse is explicit.

A second advantage is the relative simplicity and appeal of the solution. The sequence of recursive relations that define the exact solutions can be hard to interpret. In some cases, variational approaches offer explicit and more easily interpretable rules about whether or not an individual of given age and risk factors should be examined, and when an individual examined today should be examined again.

Technically, a variational solution to the optimal scheduling problem is obtained by approximating τ via a continuous function $a(t)$, termed the screening intensity function. The intensity function $a(t)$ represents the number of examinations per unit of time at age t, and its inverse $\delta(t) = 1/a(t)$ is the spacing between examination times at age t. The functional form of a can be quite general. The only condition required by this development is that it should be piecewise continuous. To find a screening schedule using a, we need first to develop an approximate objective function that depends on τ only via a; then to find the best a; and finally to derive τ from a deterministically. The last step is done using the recursive relation $\tau_0 = 0$ and

$$\tau_i = \tau_{i-1} + \delta(\tau_{i-1}), \quad i = 2, 3, \ldots .$$

This requires $a > 0$, but can also be modified to account for the case of $a = 0$ at some t. While each a generates a unique τ, the converse is not true.

To illustrate the variational approach, we consider the natural history scenario 1 of Section 6.3.4, with exponential distribution for the preclinical sojourn time, and choose a simple objective function, leading to interpretable optimality conditions. Let $\mathcal{N}(\tau)$ be the expected number of examinations and $\mathcal{D}(\tau)$ the number of cancer cases detected early among those that are bound to become symptomatic. The utility function for this section is $-k\mathcal{N}(\tau)+c\mathcal{D}(\tau)$, with exchange rate $k/c = 1/350$, leading to intervals similar to current recommendations. Call J the number of false negative results before a true positive. We begin with the case of unit sensitivity, $J = 0$. Let the hazard of preclinical disease, that is, the instantaneous probability of moving to P at age t, conditional on being in H, be

$$\lambda(t) = \frac{w_{\mathrm{hp}}(t)}{1 - W_{\mathrm{hp}}(t) - W_{\mathrm{hd}}(t)}.$$

Under regularity conditions discussed in Parmigiani (1997), the optimal screening intensity function can be determined from

$$\frac{\delta^2(t)}{2} w_{\mathrm{pc}} \left(\frac{\delta^2(t)}{2} | t \right) = \frac{k}{\lambda(t)}. \tag{6.9}$$

If (6.9) has no solution, $a = 0$ is the optimum; if (6.9) has one solution, that is the optimum; if (6.9) has two solutions, the larger a is the optimum.

When the sojourn time distribution is nearly uniform, equation (6.9) can be solved to give

$$a(t) = K \sqrt{\lambda(t)}, \tag{6.10}$$

where K is a independent of age. In this case the optimal intensity is proportional to the square root of the cause-specific failure rate λ.

Figure 6.9 compares the schedule obtained by solving (6.9) with the exact optimal screening schedule. The latter is obtained by minimizing the expected loss over the space of all examination sequences, using the recursive dynamic approach of Section 6.7.1. The intervals between examinations are approximated very closely. Based on the exact solution, the expected number of examinations is 14.45, while the probability of reaching the clinical state is 0.0634. Based on the variational solution, the expected number of examinations is 14.60, while the probability of reaching the clinical state is 0.0635.

Variational approximations can also be used to find some analytic insight into the effect of false negative results on the optimal spacing of examinations. If we define $\gamma = E(J)$, then $d\delta/d\gamma$ is approximately

$$\frac{d\delta}{d\gamma} \approx \frac{2(\delta w_{pc} + w'_{pc})}{4w_{pc} + \delta(2\gamma - 1)w'_{pc}}$$

where w'_{pc} is the partial derivative of $w_{pc}(s|t)$ with respect to s. When this is small, such as with a nearly uniform sojourn time, a unit increase in the expected number of false negative results before a true positive result leads to roughly halving the optimal spacing. The right-hand panel of Figure 6.9 also compares the screening intensity functions at $\gamma = 1$ and $\gamma = 1.25$, based on our approximation.

The results of Figure 6.9 indicate that it is optimal to increase the frequency with age. This is in disagreement with the results of Figure 6.8, which indicates that increased frequencies at young age are likely to be cost-effective. While

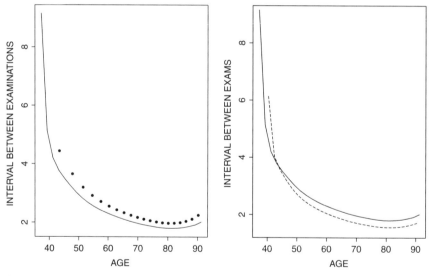

Figure 6.9 Screening schedule using the variational approximation. The left-hand panel compares the exact optimal intervals between examinations with the intervals obtained using the variational approximation. The continuous line is $1/a$ and begins at the age of first examination; the dots represent the interval between examinations in the exact case. The intervals between examinations in the exact and approximate schedule are very close, although there is some difference in the age of first examination. The right-hand panel examines the effect of test sensitivity on the variational solution, by comparing the screening intensity functions with sensitivity 1 (continuous line) and 0.8 (dotted line).

both use constant sensitivity, there are several differences between the two analyses. Of these, the principal reason for the discrepancy is likely to be the different objective functions used. The objective function of this section does not incorporate the survival implications of early detection, and does not recognize that detection at a very early stage is more important for younger women.

References

Adcock, C. J. (1997). Sample size determination: A review. *The Statistician* **46**, 261–283.

Albert, A., Gertman, P., and Louis, T. (1978). Screening for the early detection of cancer: I. The temporal natural history of a progressive disease state. *Mathematical Biosciences* **40**, 1–59.

Alexander, F. (1997). The Edinburgh randomized trial of breast cancer screening. *Journal of the National Cancer Institute Monographs* **22**, 31–35.

Allais, M. (1953). Le comportement de l'homme rationnel devant le risque: Critique des postulats et axiomes de l'école américaine. *Econometrica* **21**, 503–546.

Andersson, I. and Janzon, L. (1997). Reduced breast cancer mortality in women under age 50: Updated results from the Malmö mammographic screening program. *Journal of the National Cancer Institute Monographs* **22**, 63–67.

Angold, A. and Costello, E. J. (1993). Depressive comorbidity in children and adolescents: Empirical, theoretical, and methodological issues. *American Journal of Psychiatry* **150** (12), 1779–1791.

Anscombe, F. J. and Aumann, R. J. (1963). A definition of subjective probability. *Annals of Mathematical Statistics* **34**, 199–205.

Antman, E. M., Lau, J., Kupelnick, B., Mosteller, F., and Chalmers, T. C. (1992). A comparison of results of meta-analyses of randomized control trials and recommendations of clinical experts: Treatments for myocardial infarction. *Journal of the American Medical Association* **268**, 240–248.

Baraff, L. J., Lee, S. I., and Schriger, D. L. (1993). Outcomes of bacterial meningitis in children: A meta-analysis. *Pediatric Infectious Disease Journal* **12**, 389–394.

Barlow, R. and Proschan, F. (1965). *Mathematical Theory of Reliability*. New York: Wiley.

Barlow, R. E., Hunter, L. C., and Proschan, F. (1963). Optimum checking procedures. *Journal of the Society for Industrial and Applied Mathematics* **4**, 1078–1095.

Barnett, V. (1982). *Comparative Statistical Inference*, 2nd edn. Chichester: Wiley.

Barratt, A., Irwig, L., Glasziou, P., Cumming, R. G., Raffle, A., Hicks, N., Gray, J. A., and Guyatt, G. H. (1999). Users' guides to the medical literature: XVII. How to use guidelines and recommendations about screening. Evidence-Based Medicine Working Group. *Journal of the American Medical Association* **281**, 2029–2034.

Barry, M. J., Cantor, R., and Clarke, J. R. (1986). *Basic Diagnostic Test Evaluation*. Washington, DC: Society for Medical Decision Making.

Bayes, T. (1763). An essay towards solving a problem in the doctrine of chances. *Philosophical Transactions of the Royal Society of London* **53**, 370–418.

Beecher, H. K. (1955). The powerful placebo. *Journal of the American Medical Association* **159**, 1602–1606.

Bellman, R. E. (1957). *Dynamic Programming*. Princeton, NJ: Princeton University Press.

Benichou, J., Byrne, C., and Gail, M. H. (1997). An approach to estimating exposure-specific rates of breast cancer from a two-stage case-control study within a cohort. *Statistics in Medicine* **16**, 133–151.

Bentham, J. (1876). *An Introduction to the Principles of Morals and Legislation*. Oxford: Clarendon Press.

Berger, J. O. (1985). *Statistical Decision Theory and Bayesian Analysis*, 2nd edn. New York: Springer-Verlag.

Berger, J. O. and Berry, D. (1988). The relevance of stopping rules in statistical inference. In S. S. Gupta and J. O. Berger (eds), *Statistical Decision Theory and Related Topics IV*, pp. 29–72. Berlin: Springer-Verlag.

Berger, J. O. and Wolpert, R. L. (1988). *The Likelihood Principle*, 2nd edn. Hayward, CA: Institute of Mathematical Statistics.

Bernardo, J. M. and Smith, A. F. M. (1994). *Bayesian Theory*. New York: Wiley.

Bernoulli, D. (1738). Specimen theoriae novae de mensura sortis. *Commentarii Academiae Scientiarum Imperialis Petropolitanae* **5**, 175–192. Reprinted (1954) in *Econometrica* **22**, 23–26.

Berry, D. A. (1990). A Bayesian approach to multicenter trials and meta-analysis. In *ASA Proceedings of Biopharmaceutical Section*, pp. 1–10. Alexandria, VA: American Statistical Association.

Berry, D. A. (1996). *Statistics: A Bayesian Perspective*. Belmont, CA: Duxbury.

Berry, D. A. (1998). Benefits and risks of screening mammography for women in their forties: A statistical appraisal. *Journal of the National Cancer Institute* **90**, 1431–1439.

Berry, D. A. and Parmigiani, G. (1997). Assessing the benefits of testing for breast cancer susceptibility genes: A decision analysis. *Breast Disease.* **10**, 115–125.

Berry, D. A. and Stangl, D. K. (eds) (1996). *Bayesian Biostatistics*, Statistics: Textbooks and Monographs 151. New York: Marcel Dekker.

Berry, D. A., Parmigiani, G., Sanchez, J., Schildkraut, J., and Winer, E. (1997). Probability of carrying a mutation of breast-ovarian cancer gene *BRCA1* based on family history. *Journal of the National Cancer Institute* **89**, 227–238.

Berry, S. L. (1999). Understanding and testing for heterogeneity across 2 × 2 tables: Application to meta-analysis. *Statistics in Medicine.* **17**, 2353–2369.

Best, N. G., Cowles, M. K., and Vines, K. (1995). CODA: Convergence Diagnostics and Output Analysis Software for Gibbs Sampling Output, Version 0.30. Technical report, MRC Biostatistics Unit, University of Cambridge.

Bhat, U. N. (1984). *Elements of Applied Stochastic Processes*, 2nd edn. New York: Wiley.

Bielza, C., Müller, P., and Ríos Insua, D. (1999). Decision analysis by augmented probability simulation. *Management Science.* **45**, 995–1007.

Biggerstaff, B. J., Tweedie, R. L., and Mengersen, K. L. (1994). Passive smoking in the workplace: Classical and Bayesian meta-analyses. *International Archives of Occupational and Environmental Health* **66**, 269–277.

Billings, P. R., Kohn, M. A., de Cuevas, M., Beckwith, J., Alper, J. S., and Natowicz, M. R. (1992). Discrimination as a consequence of genetic testing. *American Journal of Human Genetics* **50**, 476–482.

Bishop, Y. M. M., Fienberg, S. E., and Holland, P. W. (1975). *Discrete Multivariate Analyses: Theory and Practice.* Cambridge, MA: MIT Press.

Bjurstam, N., Björneld, L., Duffy, S. W., Smith, T. C., Cahlin, E., Erikson, O., Hafström, L.-O., Lingaas, H., Mattsson, H., Persson, S., Rudenstam, C.-M., and Säve-Söderbergh, J. (1997). The Gothenburg breast screening trial: First results on mortality, incidence, and mode of detection for women ages 39–49 years at randomization *Cancer* **80**, 2091–2099.

Blackwood, M. A. and Weber, B. L. (1998). *BRCA1* and *BRCA2*: from molecular genetics to clinical medicine. *Journal of Clinical Oncology* **16**, 1969–1977.

Bluman, L. G., Rimer, B. K., Berry, D. A., Borstelmann, N., Iglehart, J. D., Regan, K., Schildkraut, J., and Winer, E. P. (1999). Attitudes, knowledge, and risk perceptions of women with breast and/or ovarian cancer considering testing for *BRCA1* and *BRCA2*. *Journal of Clinical Oncology* **17**, 1040–1046.

Box, G. E. P. and Draper, N. R. (1975). Robust designs. *Biometrika* **62**, 347–352.

Brookmeyer, R., Day, N. E., and Moss, S. (1986). Case-control studies for estimation of the natural history of preclinical disease from screening data. *Statistics in Medicine* **5**, 127–138.

Brophy, J. M. and Joseph, L. (1995). Placing trials in context using Bayesian analysis. GUSTO revisited by Reverend Bayes. *Journal of the American Medical Association* **273**, 871–875.

Brown, M. L. (1992). Sensitivity analysis in the cost-effectiveness of breast cancer screening. *Cancer* **69** (7 Suppl.), 1963–1967.

Brown, M. L. and Fintor, L. (1993). Cost-effectiveness of breast cancer screening: Preliminary results of a systematic review of the literature. *Breast Cancer Research and Treatment* **25**, 113–118.

Bult, J. R., Hunink, M. G. M., de Vries, J. A., and Weinstein, M. C. (1995). A fully stochastic cost-effectiveness analysis. *Medical Decision Making* **15**, 415.

Cady, B. (1996). Is axillary lymph node dissection necessary in the routine management of breast cancer? No. In V. T. DeVita, S. Hellman, and S. A. Rosenberg (eds), *Important Advances in Oncology*, pp. 251–265. Philadelphia: Lippincott-Raven.

Canadian Task Force on Preventive Health Care (2001). Preventive health care, 2001 update: Screening mammography among women aged 40–49 years at average risk of breast cancer. *Canadian Medical Association Journal* **164**, 469–476.

Carlin, B. P. and Louis, T. A. (2000). *Bayes and Empirical Bayes Methods for Data Analysis*. Boca Raton, FL: Chapman & Hall.

Carlin, B. P. and Polson, N. G. (1992). Monte Carlo Bayesian methods for discrete regression models and categorical time series. In: J. M. Bernardo, J. O. Berger, A. P. Dawid, and A. F. M. Smith (eds), *Bayesian Statistics 4*, pp. 577–586. Oxford: Oxford University Press.

Carlin, J. B. (1992). Meta-analysis for 2 × 2 tables: A Bayesian approach. *Statistics in Medicine* **11**, 141–158.

Carter, C. L., Allen, C., and Henson, D. E. (1989). Relation of tumor size, lymph node status and survival in 24,740 cases of breast cancer. *Cancer* **63**, 181–187.

Carter, R., Glasziou, P., van Oortmarssen, G., de Koning, H., Stevenson, C., Salkeld, G., and Boer, R. (1993). Cost-effectiveness of mammographic screening in Australia. *Australian Journal of Public Health* **17**, 42–50.

Chambers, I. and Altman, D. G. (1994). *Systematic Reviews*. London: BMJ Publishing Group.

Chambers, J. M. and Hastie, T. J. E. (1991). *Statistical Models in S*. Pacific Grove, CA: Wadsworth & Brooks/Cole.

Chapman, G. B. and Sonnenberg, F. B. (2000). *Decision Making in Health Care: Theory, Psychology, and Applications*. Cambridge: Cambridge University Press.

Charlton, B. G. (1996). The uses and abuses of meta-analysis. *Family Practice* **13**, 397–401.

Chib, S. and Greenberg, E. (1998). Analysis of multivariate probit models. *Biometrika* **85**, 347–361.

Clark, R. A. (1992). Economic issues in screening mammography. *American Journal of Roentgenology* **158**, 527–534.

Claus, E. B., Risch, N., and Thompson, W. D. (1990). Age of onset as an indicator of familial risk of breast cancer. *American Journal of Epidemiology* **131**, 961–972.

Clemen, R. T. and Reilly, T. (2001). *Making Hard Decisions*. Pacific Grove, CA: Duxbury.

Clyde, M. A., Müller, P., and Parmigiani, G. (1995). *Exploring Expected Utility Surfaces by Markov Chains*. Discussion paper 95-39, Institute of Statistics and Decision Sciences, Duke University. ftp://ftp.isds.duke.edu/pub/WorkingPapers/9539. ps, accessed August 2001.

Condorcet, Marquis de (1785). *Essai sur l'application de l'analyse à la probabilité des decisions rendues à la pluralité des voix*. Paris: Imprimerie Royale.

Cook, D. J., Sackett, D. L., and Spitzer, W. O. (1995). Methodologic guidelines for systematic reviews of randomized control trials in health care from the Potsdam consultation on meta-analysis. *Journal of Clinical Epidemiology* **48**(1), 167–171.

Couch, F. J., DeShano, M. L., Blackwood, M. A., Calzone, K., Stopfer, J., Campeau, L., Ganguly, A., Rebbeck, T., Weber, B. L., Jablon, L., Cobleigh, M. A., Hoskins, K., and Garber, J. E. (1997). *BRCA1* mutations in women attending clinics that evaluate the risk of breast cancer. *New England Journal of Medicine* **336**, 1409–1415.

Critchfield, G. C. and Willard, K. E. (1986). Probabilistic analysis of decision trees using monte carlo simulation. *Medical Decision Making* **6**(2), 85–92.

Crowley, P., Chalmers, I., and Keirse, M. J. N. C. (1990). The effects of corticosteroid administration before preterm delivery: An overview of the evidence from controlled trials. *British Journal of Obstetrics and Gynaecology* **97**, 11–25.

Davidson, N. E. and Abeloff, M. D. (1994). Adjuvant therapy of breast cancer. *World Journal of Surgery* **18**, 112–116.

Day, N. E. and Walter, S. D. (1984). Simplified models of screening for chronic disease: Estimation procedures from mass screening programmes. *Biometrics* **40**, 1–13.

de Finetti, B. (1937). La prévision: ses lois logiques, ses sources subjectives. *Annales de 'Institut Henri Poincaré (Paris)* **7**, 1–68.

de Koning, H. J., van Ineveld, B. M., van Oortmarssen, G. J., de Haes, J. C., Collette, H. J., Hendriks, J. H., and van der Maas, P. J. (1991). Breast cancer screening and cost-effectiveness; policy alternatives, quality of life considerations and the possible impact of uncertain factors. *International Journal of Cancer* **49**, 531–537.

Decision Analysis Society (2001). Decision tree & influence diagram software. http://faculty.fuqua.duke.edu/daweb/dasw6.htm, accessed July 2001.

Deeks, J., Glanville, J., and Sheldon, T. (1997). Undertaking systematic reviews of research on effectiveness: CRD guidelines for those carrying out or commissioning reviews. Report 4, Centre for Reviews and Dissemination, York.

DeGroot, M. H. (1970). *Optimal Statistical Decisions*. New York: McGraw-Hill.

DeGroot, M. H. (1984). Changes in utility as information. *Theory and Decision* **17**, 287–303.

DeGroot, M. H. (1986). *Probability and Statistics*. Reading, MA: Addison-Wesley.

DerSimonian, R. and Laird, N. (1986). Meta-analysis in clinical trials. *Controlled Clinical Trials* **7**, 177–188.

Diaconis, P. and Ylvisaker, D. (1979). Conjugate priors for exponential families. *Annals of Statistics* **7**, 269–281.

Diebolt, J. and Robert, C. P. (1994). Estimation of finite mixture distributions through Bayesian sampling. *Journal of the Royal Statistical Society B* **56**, 363–375.

Dominici, F. and Parmigiani, G. (2000). Combining studies with continuous and dichotomous responses: A latent variables approach. In: D.K Stangl and D. A. Berry (eds), *Meta-analysis in Medicine and Health Policy*, pp. 105–126. New York: Marcel Dekker.

Dominici, F., Parmigiani, G., Wolpert, R. L., and Hasselblad, V. (1999). Meta-analysis of migraine headache treatments: Combining information from heterogeneous designs. *Journal of the American Statistical Association* **94**, 16–28.

Doubilet, P., Begg, C. B., Weinstein, M. C., Braun, P., and McNeil, B. J. (1985). Probabilistic sensitivity analysis using Monte Carlo simulation. a practical approach. *Medical Decision Making* **5**, 157–177.

Doubilet, P., Weinstein, M. C., and McNeil, B. J. (1986). Use and misuse of the term 'cost effective' in medicine. *New England Journal of Medicine* **314**(4), 253–256.

Draper, D. (1995). Assessment and propagation of model uncertainty (with discussion). *Journal of the Royal Statistical Society B* **57**, 45–97.

Drum, D. E. and Christacapoulos, J. S. (1972). Hepatic scintigraphy in clinical decision making. *Journal of Nuclear Medicine* **13**, 908–915.

Duda, R. O., Hart, P. E., and Stork, D. G. (2001). *Pattern Classification*. New York: Wiley.

DuMouchel, W. H. (1990). Bayesian metaanalysis. In D. A. Berry (ed.), *Statistical Methodology in the Pharmaceutical Sciences*, Statistics: Textbooks and Monographs 104, pp. 509–529. New York: Marcel Dekker.

DuMouchel, W. H. and Harris, J. E. (1983). Bayes methods for combining the results of cancer studies in humans and other species (with discussion). *Journal of the American Statistical Association* **78**, 293–315.

Duval, S. and Tweedie, R. L. (2000). Trim and fill: A simple funnel-plot-based method of testing and adjusting for publication bias in meta-analysis. *Biometrics* **56**, 455–463.

Early Breast Cancer Trialists Collaborative Group (1990). *Treatment of Early Breast Cancer, Vol. 1: Worldwide Evidence 1985–1990.* Oxford: Oxford University Press.

Early Breast Cancer Trialists Collaborative Group (1992). Systemic treatment of early breast cancer by hormonal, cytotoxic, or immune therapy. *Lancet* **339**, 1–15, 71–85.

Early Breast Cancer Trialists Collaborative Group (1995). Effects of radiotherapy and surgery in early breast cancer. *New England Journal of Medicine* **333**, 1444–1455.

Early Breast Cancer Trialists Collaborative Group (1998a). Tamoxifen for early breast cancer: An overview of the randomised trials. *Lancet* **351**, 1451–1467.

Early Breast Cancer Trialists Collaborative Group (1998b). Poly-chemo-therapy for early breast cancer: An overview of the randomised trials. *Lancet* **352**, 930–942.

Easton, D. F., Ford, D., and Bishop, D. T. (1995). Breast and ovarian cancer incidence in *BRCA1*-mutation carriers. *American Journal of Human Genetics* **56**, 265–271.

Eddy, D. M. (1980). *Screening for Cancer: Theory Analysis and Design.* Englewood Cliffs, NJ: Prentice Hall.

Eddy, D. M. (1983). A mathematical model for timing repeated medical tests. *Medical Decision Making* **3**, 34–62.

Eddy, D. M., Hasselblad, V., McGivney, W., and Hendee, W. (1989). The value of mammography screening in women under age 50 years. *Journal of the American Medical Association* **259**, 1512–1519.

Eddy, D. M., Hasselblad, V., and Shachter, R. (1990). A Bayesian method for synthesizing evidence. The confidence profile method. *International Journal of Technology Assessment in Health Care* **6**, 31–55.

Edwards, W., Lindman, H., and Savage, L. J. (1963). Bayesian statistical inference for psychological research. *Psychological Bulletin* **20**, 193–242.

Elixhauser, A. (1991). Costs of breast cancer and the cost-effectiveness of breast cancer screening *International Journal of Technology Assessment in Health Care* **7**, 604–615.

Elmore, J. G., Barton, M. B., Moceri, V. M., Polk, S., Arena, P. J., and Fletcher, S. W. (1998). Ten-year risk of false positive screening mammograms and clinical breast examinations. *New England Journal of Medicine* **338**, 1089–1096.

Elston, R. and Stewart, J. (1971). A general model for the genetic analysis of pedigree data. *Human Heredity* **21**, 523–542.

Epstein, R. J. (1995). Routine or delayed axillary dissection for primary breast cancer? *European Journal of Cancer* **31A**, 1570–1573.

Erkanli, A., Soyer, R., and Angold, A. (1998). Optimal Bayesian two-phase designs. *Journal of Statistical Planning and Inference* **66**, 175–191.

Erwin, E. (1984). Establishing causal connections: Meta-analysis and psychotherapy. *Midwest Studies in Philosophy* **9**, 421–436.

Etzioni, R. D. and Shen, Y. (1997). Estimating asymptomatic duration in cancer: The AIDS connection. *Statistics in Medicine* **16**, 627–644.

Evidence-Based Medicine Working Group (1992). Evidence-based medicine: A new approach to teaching the practice of medicine. *Journal of the American Medical Association* **268**, 2420–2425.

Fagan, T. J. (1975). Nomogram for Bayes' formula. *New England Journal of Medicine* **293**, 257.

Fanshel, S. and Bush, J. (1970). A health status index and its application to the health services outcomes. *Operations Research* **18**, 1021–1066.

Feig, S. A. and Hendrick, R. E. (1997). Radiation risk from screening mammography of women aged 40–49 years. *Journal of the National Cancer Institute Monographs* **22**, 119–124.

Feinstein, A. R. (1977). Clinical biostatistics XXXIX. The haze of Bayes, the aerial palaces of decision analysis, and the computerized ouija board. *Clinical Pharmacology* **21**, 482–496.

Fishburn, P. C. (1970). *Utility Theory for Decision Making.* New York: Wiley.

Fishburn, P. C. (1989). Retrospective on the utility theory of von Neumann and Morgenstern. *Journal of Risk and Uncertainty* **2**, 127–158.

Fisher, B., Redmond, C., and Fisher, E. R. (1985). Ten-year results of a randomized clinical trial comparing radical mastectomy and total mastectomy with or without radiation. *New England Journal of Medicine* **312**, 674–681.

Fisher, B., Dignam, J., Bryant, J., DeCillis, A., Wickerham, D. L., and Wolmark, N. (1996). Five versus more than five years of tamoxifen therapy for breast cancer patients with negative lymph nodes and estrogen receptor-positive tumors. *Journal of the National Cancer Institute* **88**, 1529–1542.

Fisher, R. A. (1925). Theory of statistical estimation. *Proceedings of the Cambridge Philosophical Society* **22**, 700–725.

Flehinger, B. J. and Kimmel, M. (1994). Early lung cancer detection studies. In N. Lange *et al.* (eds), *Case Studies in Biometry*, pp. 301–321. New York: Wiley.

Ford, D. and Easton, D. F. (1995). The genetics of breast and ovarian cancer. *British Journal of Cancer* **72**, 805–812.

Ford, D., Easton, D. F., Stratton, M., Narod, S., Goldgar, D., Devilee, P., Bishop, D. T., *et al.* (1998). Genetic heterogeneity and penetrance analysis of the *BRCA1* and *BRCA2* genes in breast cancer families. *American Journal of Human Genetics* **62**, 676–689.

Forrester, J. W. (1961). *Industrial Dynamics.* Cambridge, MA: MIT Press.

Frank, T., Manley, S., Olopade, O., Cummings, S., Garber, J., Bernhardt, B., Antman, K., *et al.* (1998). Sequence analysis of *BRCA1* and *BRCA2*:

Correlation of mutations with family history and ovarian cancer risk. *Journal of Clinical Oncology* **16**, 2417–2425.

French, S. (1986). *Decision Theory: An Introduction to the Mathematics of Rationality*. Chichester: Ellis Horwood.

Frisell, J. and Lidbrink, E. (1997). The Stockholm mammographic screening trial: Risks and benefits in age group 40–49 years. *Journal of the National Cancer Institute Monographs* **22**, 49–51.

Futreal, P. A., Liu, Q., Shattuck-Eidens, D., Cochran, C., Harshman, K., Tavtigian, S., Bennett, L. M., *et al.* (1994). *BRCA1* mutations in primary breast and ovarian carcinomas. *Science* **226**, 120–122.

Gail, M. H. and Benichou, J. (1994). Validation studies on a model for breast cancer risk (editorial). *Journal of the National Cancer Institute* **86**, 573–575.

Gail, M. H., Brinton, L. A., Byar, D. P., Corle, D. K., Green, S. B., Schairer, C., and Mulvihill, J. J. (1989). Projecting individualized probabilities of developing breast cancer for white females who are being examined annually. *Journal of the National Cancer Institute* **81**, 1879–1886.

Gail, M. H., Costantino, J. P., Bryant, J., Croyle, R., Freedman, L., Helzlsouer, K., and Vogel, V. (1999). Weighing the risks and benefits of tamoxifen treatment for preventing breast cancer. *Journal of the National Cancer Institute* **91**, 1829–1846.

Gamerman, D. (1997). *Markov Chain Monte Carlo: Stochastic Simulation for Bayesian Inference*. London: Chapman & Hall.

Gardiner, J., Hogan, A., Holmes-Rovner, M., Rovner, D., Griffith, L., and Kupersmith, J. (1995). Confidence intervals for cost-effectiveness ratios. *Medical Decision Making* **15**, 254–263.

Gelber, R. D. and Goldhirsch, A. (1991). Meta-analysis: the fashion of summing-up evidence. Part I. Rationale and conduct. *Annals of Oncology* **2**, 461–468.

Gelber, R. D., Goldhirsch, A., and Cavalli, F. (1991). Quality-of-life-adjusted evaluation of adjuvant therapies for operable breast cancer. *Annals of Internal Medicine* **114**, 621–628.

Gelfand, A. E. and Smith, A. F. M. (1990). Sampling-based approaches to calculating marginal densities. *Journal of the American Statistical Association* **85**, 398–409.

Gelman, A., Carlin, J., Stern, H., and Rubin, D. (1995). *Bayesian Data Analysis*. London: Chapman & Hall.

Gelman, A., Roberts, G. O., and Gilks, W. R. (1996). Efficient Metropolis jumping rules. In J. M. Bernardo, J. O. Berger, A. P. Dawid, and A. F. M. Smith (eds.), *Bayesian Statistics 5*, pp. 599–607, Oxford: Oxford University Press.

Geman, S. and Geman, D. (1984). Stochastic relaxation, Gibbs distributions, and the Bayesian restoration of images. *IEEE Transactions Pattern Analysis and Machine Intelligence* **6**, 721–741.

Gentle, J. E. (1998). *Random Number Generation and Monte Carlo Methods.* New York: Springer-Verlag.

Gilks, W. R., Richardson, S., and Spiegelhalter, D. J. (eds) (1996). *Markov Chain Monte Carlo in Practice.* London: Chapman & Hall.

Gilks, W. R., Thomas, A., and Spiegelhalter, D. J. (1994). A language and program for complex Bayesian modelling. *The Statistician* **43**, 169–177.

Givens, G. H., Smith, D. D., and Tweedie, R. L. (1997). Bayesian data-augmented meta-analysis that account for publication bias issues exemplified in the passive smoking debate. *Statistical Science* **12**, 221–250.

Glass, G. V. (1976). Primary, secondary and meta-analysis of research. *Educational Researcher* **5**, 3–8.

Gold, M. R., Siegel, J. E., Russell, L. B., and Weinstein, M. C. (eds) (1996). *Cost-effectiveness in Health and Medicine.* Oxford: Oxford University Press.

Goodman, K. W. (1998). Meta-analysis: Conceptual, ethical and policy issues. In K. W. Goodman (ed.), *Ethics, Computing and Medicine*, pp. 139–167. Cambridge: Cambridge University Press.

Goodman, S. N. (1999a). Towards evidence-based medical statistics. 1: The *p*-value fallacy. *Annals of Internal Medicine* **130**, 995–1004.

Goodman, S. N. (1999b). Towards evidence-based medical statistics. 2: The Bayes factor. *Annals of Internal Medicine* **130**, 1005–1013.

Gordis, L., Berry, D., Chu, S., Fajardo, L., Hoel, D., Laufmann, L., Rufenbarger, C., Scott, J., Sullivan, D., Wasson, J., Westhoff, C., and Zern, R. (1997). Breast cancer screening for women ages 40–49. *Journal of the National Cancer Institute* **89**, 1015–1026.

Gorry, G. A. and Barnett, G. O. (1968). Sequential diagnosis by computer. *Journal of the American Medical Association* **205**, 849–854.

Gorry, G. A., Pauker, S. G., and Schwartz, W. B. (1978). The diagnostic importance of normal findings. *New England Journal of Medicine* **298**, 486–489.

Goslin, R. E., Gray, R. N., McCrory, D. C., Penzien, D., Rains, J., and Hasselblad, V. (1998a). Behavioral and physical treatments for migraine. Evidence report, Agency for Health Care Policy and Research, Bethesda, MD.

Goslin, R. E., Gray, R. N., McCrory, D. C., Tulsky, J., and Hasselblad, V. (1998b). Drug treatments for prevention of migraine. Evidence report, Agency for Health Care Policy and Research, Bethesda, MD.

Habbema, J. D. F., van Oortmarssen, G. J., Lubbe, J. T. N., and van der Maas, P. J. (1985). The MISCAN simulation program for the evaluation of screening for disease. *Computer Methods & Programs in Biomedicine* **20**, 79–93.

Haines, L. M. (1987). The application of the annealing algorithm to the construction of exact optimal designs for linear-regression models. *Technometrics* **29**, 439–447.

Halsted, W. S. (1894–1895). The result of operations for the cure of the breast performed at the Johns Hopkins Hospital from June 1889 to January 1894. *Johns Hopkins Hospital Bulletin* **4**, 279.

Harris, J. R. (1996). *Breast Diseases*. Philadelphia: Lippincott.

Hartge, P., Struewing, J. P., Wacholder, S., Brody, L. C., and Tucker, M. A. (1999). The prevalence of common *BRCA1* and *BRCA2* mutations among Ashkenazi Jews. *American Journal of Human Genetics* **64**, 963–970.

Hartigan, J. A. (1983). *Bayes Theory*. Berlin: Springer-Verlag.

Hartmann, L. C., Schaid, D. J., Woods, J. E., Crotty, T. P., Myers, J. L., Arnold, P. G., Petty, P. M., Sellers, T. A., Johnson, J. L., McDonnell, S. K., Frost, M. H., and Jenkins, R. B. (1999). Efficacy of bilateral prophylactic mastectomy in women with a family history of breast cancer. *New England Journal of Medicine* **340**, 77–84.

Hasselblad, V. (1998). Meta-analysis of multitreatment studies. *Medical Decision Making* **18**, 37–43.

Hasselblad, V. and Hedges, L. V. (1995). Meta-analysis of screening and diagnostic tests. *Psychological Bulletin* **117**, 167–178.

Hasselblad, V. and McCrory, D. C. (1995). Meta-analytic tools for medical decision making: A practical guide. *Medical Decision Making* **15**, 81–96.

Hastings, W. K. (1970). Monte Carlo sampling methods using Markov chains and their applications. *Biometrika* **57**, 97–109.

Hedges, L. V. and Olkin, I. (1985). *Statistical Methods for Meta-analysis*. New York: Academic Press.

Heitjan, D. F., Moskowitz, A. J., and Whang, W. (1999). Bayesian estimation of cost-effectiveness ratios from clinical trials. *Health Economics* **8**(3), 191–201.

Hendrick, R., Smith, R., Rutledge, J. I., and Smart, C. (1997). Benefit of screening mammography in women aged 40–49: A new meta-analysis of randomized controlled trials. *Journal of the National Cancer Institute Monographs* **22**, 87–92.

Hillner, B. E. and Smith, T. J. (1991). Efficacy and cost effectiveness of adjuvant chemotherapy in women with node-negative breast cancer. *New England Journal of Medicine* **324**(3), 160–168.

Hu, P. and Zelen, M. (1997). Planning clinical trials to evaluate early detection programmes. *Biometrika* **84**, 817–829.

Hui, S. L. and Walter, S. D. (1980). Estimating the error rates of diagnostic tests. *Biometrics* **36**, 167–171.

Iglehart, J. D., Miron, A., Rimer, B. K., Winer, E. P., Berry, D. A., and Schildkraut, J. (1998). Overestimation of hereditary breast cancer risk. *Annals of Surgery* **228**, 375–384.

Irwig, L., Glasziou, P., Barratt, A., and Salkeld, G. (1997). Review of the evidence about the value of mammographic screening in 40–49 year old women. http://www.nbcc.org.au/pages/info/resource/nbccpubs/scr4049/contents.htm, accessed August 2001.

Iversen, Jr, E. S., Parmigiani, G., Berry, D., and Schildkraut, J. (1997). A Markov chain Monte Carlo approach to survival models with detailed family history. *Genetic Epidemiology* **14**, 531.

Iversen, Jr, E. S., Parmigiani, G., Berry, D. A., and Schildkraut, J. (2000). Genetic susceptibility and survival: Application to breast cancer. *Journal of the American Statistical Association* **95**, 28–42.

Jadad, A. R. and McQuay, H. J. (1996). Meta-analyses to evaluate analgesic interventions: A systematic qualitative review of their methodology. *Journal of Clinical Epidemiology* **49**, 235–243.

Jeffreys, H. (1961). *Theory of Probability.* London: Oxford University Press.

Johansen, H., Kaae, S., and Scheiodt, T. (1990). Simple mastectomy with postoperative irradiation vs. extended radical mastectomy in breast cancer. *Acta Oncologica* **29**, 709–715.

Jorland, G. (1987). The Saint Petersburg paradox 1713–1937. In L. Krüges, L. J. Daston and M. Heidelberges (eds), *The Probabilistic Revolution: Volume 1, Ideas in History*, pp. 157–190. Cambridge, MA: MIT Press.

Kadane, J. B. (1996). *Bayesian Methods and Ethics in a Clinical Trial.* New York: Wiley.

Kadane, J. B. and Chuang, D. T. (1978). Stable decision problems. *Annals of Statistics* **6**, 1095–1110.

Kadane, J. B., Schervish, M. J., and Seidenfeld, T. (1999). *Rethinking the Foundations of Statistics.* Cambridge: Cambridge University Press.

Kahneman, D., Slovic, P., and Tversky, A. (1982). *Judgment under Uncertainty: Heuristics and Biases.* Cambridge: Cambridge University Press.

Kakuda, J. T., Stuntz, M., Trivedi, V., Klein, S. R., and Vargas, H. I. (1999). Objective assessment of axillary morbidity in breast cancer treatment. *American Surgeon* **65**, 995–998.

Kaplan, R. (1995). Utility assessment for estimating quality-adjusted life years. In: F. Sloan (ed.), *Valuing Health Care*, pp. 157–190. Cambridge: Cambridge University Press.

Keller, J. B. (1974). Optimum checking schedules for systems subject to random failure. *Management Science* **21**, 256–260.

Kerlikowske, K., Grady, D., Barclay, J., Frankel, S. D., Ominsky, S. H., Sickles, E. A., and Ernster, V. (1998). Variability and accuracy in mammographic interpretation using the American College of Radiology breast imaging reporting and data system. *Journal of the National Cancer Institute* **90**, 1801–1809.

Keynes, G. (1936). Conservative treatment of cancer of the breast. *British Medical Journal* **2**, 643–647.

Kirch, R. L. A. and Klein, M. (1974). Examination schedules for breast cancer. *Cancer* **33**, 1444–1450.

Kleijnen, J. P. C. (1997). Sensitivity analysis and related analyses: A review of some statistical techniques. *Journal of Statistical Computation and Simulation* **57**, 111–142.

Kleijnen, J. P. C. and Helton, J. C. (1999). Statistical analyses of scatterplots to identify important factors in large-scale simulations, 1: Review and comparison of techniques. *Reliability Engineering & System Safety* **65**, 147–185.

Knox, E. (1973). A simulation system for screening procedures. In G. McLachlan (ed.), *The Future and Present Indicatives. Problems and Progress in Medical Care*, pp. 19–55. London: Oxford University Press.

Knuth, D. E. (1997). *The Art of Computer Programming*, Vol. 2. Reading, MA: Addison-Wesley.

Kreps, D. M. (1988). *Notes on the Theory of Choice*. Boulder, CO: Westview Press.

L'Abbé, K. A., Detsky, A. S., and O'Rourke, K. (1987). Meta-analysis in clinical research. *Annals of Internal Medicine* **107**, 224–233.

Lambert, P. C., Abrams, K. R., Sansó, B., and Jones, D. R. (1997). Synthesis of incomplete data using Bayesian hierarchical models: An illustration based on data describing survival from neuroblastoma. Technical report, University of Leicester.

Lange, K. (1986). Approximate confidence intervals for risk prediction in genetic counseling. *American Journal of Human Genetics* **38**, 681–687.

Lashner, B. A., Hanauer, S. B., and Silverstein, M. D. (1988). Optimal timing of colonoscopy to screen for cancer in ulcerative colitis. *Annals of Internal Medicine* **108**, 274–278.

Leal, S. M. and Ott, J. (1995). Variability of genotype-specific penetrance probabilities in the calculation of risk support intervals. *Genetic Epidemiology*, **12**, 859–862.

Lee, S. J. and Zelen, M. (1998). Scheduling periodic examinations for the early detection of disease: Applications to breast cancer. *Journal of the American Statistical Association* **93**, 1271–1281.

Lenert, L. and Kaplan, R. M. (2000). Validity and interpretation of preference-based measures of health-related quality of life. *Medical Care* **38** (9 Suppl.), II138–150.

Lin, P. P., Allison, D. C., Wainstock, J., Miller, K. D., Dooley, W. C., Friedman, N., *et al.* (1993). Impact of axillary lymph node dissection on the therapy of breast cancer patients. *Journal of Clinical Oncology* **11**, 1536–1544.

Lincoln, T. L. and Weiss, G. H. (1964). A statistical evaluation of recurrent medical evaluations. *Operations Research* **12**, 187–205.

Lindfors, K. K. and Rosenquist, C. J. (1995). The cost-effectiveness of mammographic screening strategies. *Journal of American Medical Association* **274**, 881–884.

Lindley, D. V. (1965). *Introduction to Probability and Statistics from a Bayesian Viewpoint*. Cambridge: Cambridge University Press.

Lindley, D. V. (1985). *Making Decisions*, 2nd edn. New York: Wiley.

Lindley, D. V. (1997). The choice of sample size. *The Statistician* **46**, 129–138.

Lindley, D. V. and Smith, A. F. M. (1972). Bayes estimates for the linear model (with discussion). *Journal of the Royal Statistical Society B* **34**, 1–41.

Lipton, R. B. and Stewart, W. F. (1993). Migraine in the United States: A review of epidemiology and health care use. *Neurology* **43**, S6–S10.

Lusted, L. B. (1968). *Introduction to Medical Decision Making*. Springfield, IL: Thomas.

Lynch, H. T., Albano, W. A., Heieck, J. J., Mulcahy, G. M., Lynch, J. F., Layton, M. A., and Danes, B. S. (1984). Genetics, biomarkers and control of breast cancer: A review. *Cancer Genetics and Cytogenetics* **13**, 43–92.

Manning, W. G., Fryback, D. G., and Weinstein, M. C. (1996). Reflecting uncertainty in cost-effectiveness analyses. In: M. R. Gold, J. E. Siegel, L. B. Russell, and M. C. Lleinstein (eds), *Cost-effectiveness in Health and Medicine*. Oxford: Oxford University Press.

Manton, K. G. and Stallard, E. (1988). *Chronic Disease Modelling*. London: Charles Griffin.

Matchar, D. B. (1998). The value of stroke prevention and treatment. *Neurology* **51** (3 Suppl.), S31–5.

Matchar, D. B. and Samsa, G. P. (1999). Using outcomes data to identify best medical practice: the role of policy models. *Hepatology* **29**(6 Suppl.), 36S–39S.

Matchar, D. B., Duncan, P., Samsa, G., *et al.* (1993). The stroke prevention patient outcomes research team: goals and methods. *Stroke* **24**(12), 2135–2142.

Matchar, D. B., Samsa, G. P., Matthews, J. R., Ancukiewicz, M., Parmigiani, G., Hasselblad, V., Wolf, P. A., D'Agostino, R. B., and Lipscomb, J. (1997). The Stroke Prevention Policy Model (SPPM): Linking evidence and clinical decisions. *Annals of Internal Medicine* **127**(8S), 704–711.

McNeil, B., Weichselbaum, R., and Pauker, S. (1981). Speech and survival: Tradeoffs between quality and quantity of life in laryngeal cancer. *New England Journal of Medicine* **305**, 982–987.

Meissen, G. J., Mastromauro, C. A., Kiely, D. K., McNamara, D. S., and Myers, R. H. (1991). Understanding the decision to take the predictive test for Huntington disease. *American Journal of Medical Genetics* **39**(4), 404–410.

Meng, X.-L. (1994). Posterior predictive p-values. *Annals of Statistics* **22**, 1142–1160.

Metropolis, N. (1987). The beginning of the Monte Carlo method. *Los Alamos Science* **15**, 125–130.

Metropolis, N., Rosenbluth, A. W., Rosenbluth, M. N., Teller, A. H., and Teller, E. (1953). Equations of state calculations by fast computing machine. *Journal of Chemical Physics* **21**, 1087–1091.

Meulders, M., Gelman, A., van Mechelen, I., and de Boeck, P. (1998). Generalizing the probability matrix decomposition: An example of Bayesian model checking and model expansion. In: J. J. Hox and E. D. de Leeuw (eds), *Assumptions, Robustness, and Estimation Methods in Multivariate Modeling*, pp. 1–19. Amesterdam: TT-Publikaties.

Miki, Y., Swenson, J., Shattuck-Eidens, D., Futreal, P. A., Harshman, K., Tavtigian, S., et al. (1994). A strong candidate for the breast and ovarian cancer susceptibility: Gene *BRCA1*. *Science* **266**, 66–71.

Miller, A., To, T., Baines, C., and Wall, C. (1997). The Canadian National Breast Screening Study: Update on breast cancer mortality. *Journal of the National Cancer Institute Monographs* **22**, 37–41.

Moher, D. and Olkin, I. (1995). Meta-analysis of randomized controlled trials: A concern for standards. *Journal of the American Medical Association* **48**, 1952–1964.

Moher, D., Cook, D. J., Eastwood, S., Olkin, I., Rennie, D., Stroup, D. F., and QUOROM Group (1999). Improving the quality of reports of meta-analyses of randomised controlled trials. *Lancet* **354**, 1896–1900.

Moolgavkar, S. H., Stevens, R. G., and Lee, J. A. H. (1979). Effect of age on incidence of breast cancer in females. *Journal of the National Cancer Institute* **62**, 493–501.

Moore, M. P. and Kinne, D. W. (1996). Is axillary lymph node dissection necessary in the routine management of breast cancer? Yes. In V. T. DeVita, S. Hellman, and S. A. Rosenberg (eds), Advances in Oncology, pp. 245–250. Philadelphia: Lippincott-Raven.

Morris, A., Morris, R., Wilson, J., White, J., Steinberg, S., Okunieff, P., et al. (1997). Breast conserving therapy vs. mastectomy in early stage breast cancer: A meta-analysis of ten years' survival. *Cancer Journal* **3**, 6–12.

Morris, C. N. and Normand, S.-L. (1992). Hierarchical models for combining information and for meta-analysis. In: J. M. Bernardo, J. O. Beger, A. P. David and A. F. M. Smith (eds), *Bayesian Statistics 4*, 321–344. Oxford: Oxford University Press.

Morrison, A. (1989). Review of evidence on the early detection and treatment of breast cancer. *Cancer* **64**, 2651–2656.

Moskowitz, M. (1986). Breast cancer: Age-specific growth rates and screening strategies. *Radiology* **161**, 37–41.

Mullahi, J. and Manning, W. G. (1996). Statistical issues in cost-effectiveness analyses. In: M. R. Gold, J. E. Siegel, L. B. Russell, and M. C. Lkinstein (eds), *Cost-effectiveness in Health and Medicine*. Oxford: Oxford University Press.

Müller, P. (1998). Simulation based optimal design. In: J. M. Bernardo, J. O. Berger, A. P. David and A. F. M. Smith (eds), *Bayesian Statistics 6*. Oxford: Oxford University Press.

Müller, P. and Parmigiani, G. (1995). Optimal design via curve fitting of Monte Carlo experiments. *Journal of the American Statistical Association* **90**, 1322–1330.

Murphy, E. A. and Mutalik, G. S. (1969). The application of Bayesian methods in genetic counseling. *Human Heredity* **19**, 126–151.

Mushlin, A. I. and Fintor, L. (1992). Is screening for breast cancer cost-effective? *Cancer* **69** (7 Suppl.), 1957–1962.

Mushlin, A. I., Kouides, R. W., and Shapiro, D. E. (1998). Estimating the accuracy of screening mammography: A meta-analysis. *American Journal of Preventive Medicine* **14**(2), 143–153.

Myriad Genetics (2001). Mutation prevalence tables http://www.myriad.com/ med/brac/mutptables.html, accessed August 2001.

Narod, S. A., Goldgar, D., Cannon-Albright, L., Weber, B., Moslehi, R., Ives, E., Lenoir, G., and Lynch, H. (1995). Risk modifiers in carriers of *BRCA1* mutations. *International Journal of Cancer* **64**, 394–398.

National Action Plan on Breast Cancer and American Society of Clinical Oncology (1997). Hereditary susceptibility to breast and ovarian cancer: An outline of the basic fundamental knowledge needed by all health care professionals. http://www.4woman.gov/napbc/napbc/hsedcurr.htm, accessed August 2001.

National Cancer Institute (1997). Surveillance, epidemiology, and end results http://www-seer.ims.nci.nih.gov, accessed August 2001.

National Institutes of Health Consensus Development Conference (1991). Early stage breast cancer. *Journal of the American Medical Association* **265** (3), 391–395.

Neal, R. M. (1996). *Bayesian Learning for Neural Networks*. New York: Springer-Verlag.

Newman, B., Millikan, R. C., and King, M.-C. (1997). Genetic epidemiology of breast and ovarian cancers. *Epidemiologic Reviews* **19**, 69–79.

Neyman, J. and Pearson, E. S. (1933). On the problem of the most efficient test of statistical hypotheses. *Philosophical Transactions of the Royal Society* **231**, 286–337.

Niederreiter, H. (1992). *Random Number Generation and Quasi-Monte Carlo Methods*. Philadelphia: Society of Industrial and Applied Mathematics.

Nord, E. (1992). Methods for quality adjustment of life years. *Social Science and Medicine* **34**(5), 559–569.

Normand, S.-L.T. (1995). Meta-analysis software: A comparative review. *American Statistician* **49**, 298–309.

Oddoux, C., Struewing, J. P., Clayton, C. M., Neuhausen, S., Brody, L. C., Kaback, M., Haas, B., Norton, L., Borgen, P., Jhanwar, S., Goldgar, D., Ostrer, H., and Offit, K. (1996). The carrier frequency of the *BRCA2*

6174delT mutation among Ashkenazi Jewish individuals is approximately 1%. *Nature Genetics* **14**, 188–190.

Offit, K. and Brown, K. (1994). Quantitating familial cancer risk: A resource for clinical oncologists. *Journal of Clinical Oncology* **12**, 1724–1736.

O'Hagan, A. (1994). *Kendall's Advanced Theory of Statistics. Volume 2B: Bayesian Inference*. London: Edward Arnold.

Olkin, I. (1995a). Meta-analysis: Reconciling the results of independent studies. *Statistics in Medicine* **14**, 457–472.

Olkin, I. (1995b). Statistical and theoretical considerations in meta-analysis. *Journal of Clinical Epidemiology* **274**(24), 133–146.

Orr, R. K., Col, N. F., and Kuntz, K. M. (1999). A cost-effectiveness analysis of axillary node dissection in postmenopausal women with estrogen receptor-positive breast cancer and clinically negative axillary nodes. *Surgery* **126**, 568–576.

Pallay, A. and Berry, S. M. (1999). A decision analysis for an end of Phase II go/stop decision. *Drug Information Journal* **33**, 821–833.

Parmigiani, G. (1993). On optimal screening ages. *Journal of the American Statistical Association* **88**, 622–628.

Parmigiani, G. (1997). Timing medical examinations via intensity functions. *Biometrika* **84**, 803–816.

Parmigiani, G. (1998). Decision models in screening for breast cancer. In: J. M. Bernardo, J. O. Berger, A. P. David and A. F. M. Smith (eds), *Bayesian Statistics 6*. Oxford: Oxford University Press.

Parmigiani, G. (2001). Uncertainty and the value of diagnostic information. Technical report, Johns Hopkins School of Medicine.

Parmigiani, G. and Berry, D. A. (1994). Applications of Lindley information measure to the design of clinical experiments. In A. F. M. Smith and P. R. Freeman (eds), *Aspects of Uncertainty. A Tribute to D. V. Lindley*, pp. 351–362. Chichester: Wiley.

Parmigiani, G. and Kamlet, M. (1993). Cost–utility analysis of alternative strategies in screening for breast cancer. In: C. Gatsonis, J. S. Hodges, R. E. Kass, and N. D. Singpurwalla (eds), *Case Studies in Bayesian Statistics*, Lecture Notes in Statistics 83, pp. 390–402. New York: Springer-Verlag.

Parmigiani, G. and Skates, S. (2001). Estimating the age of onset of detectable asymptomatic cancer. *Mathematical and Computer Modeling* **33**, 1347–1360.

Parmigiani, G., Berry, D. A., and Aguilar, O. (1998). Determining carrier probabilities for breast cancer susceptibility genes *BRCA1* and *BRCA2*. *American Journal of Human Genetics* **62**, 145–158.

Parmigiani, G., Samsa, G. P., Ancukiewicz, M., Lipscomb, J., Hasselblad, V., and Matchar, D. B. (1997). Assessing uncertainty in cost-effectiveness analyses: Application to a complex decision model. *Medical Decision Making* **17**, 390–401.

Parmigiani, G., Berry, D. A., Winer, E. P., Tebaldi, C., Iglehart, J. D., and Prosnitz, L. (1999). Is axillary lymph node dissection indicated for early stage breast cancer – a decision analysis. *Journal of Clinical Oncology* **17**(5), 1465–1473.

Patrick, D. L. and Erickson, P. (1993). *Health Status and Health Policy: Allocating Resources to Health Care.* New York: Oxford University Press.

Pauker, S. G. and Pauker, S. P. (1987). Prescriptive models to support decision making in genetics. *Birth Defects* **23**, 279–296.

Peer, P. G. M., van Dijck, J. A. A. M., Hendriks, J., Holland, R., and Verbeek, A. L. M. (1993). Age-dependent growth rate of primary breast cancer. *Cancer* **71**, 3547–3551.

Peer, P. G. M., Verbeek, A. L. M., and Straatman, H. (1995). Sample size determination for a trial of breast cancer screening under age 50: Population versus case mortality approach. *Journal of Medical Screening* **2**, 90–93.

Peer, P. G. M., Verbeek, A., Straatman, H., Hendriks, J., and Holland, R. (1996a). Age-specific sensitivities of mammographic screening for breast cancer. *Breast Cancer Research and Treatment* **38**, 153–160.

Peer, P. G. M., Verbeek, A. L., Mravunac, M., Hendriks, J. H., and Holland, R. (1996b). Prognosis of younger and older patients with early breast cancer. *British Journal of Cancer* **73**, 382–385.

Perloff, M., Norton, L., Korzun, A., Wood, W., Carey, R., Gottlieb, A., *et al.* (1996). Postsurgical adjuvant chemotherapy of stage II breast carcinoma with or without crossover to a non-cross-resistant regimen: A cancer and leukemia group B study. *Journal of Clinical Oncology* **14**, 1589–1598.

Pham-Gia, T. and Turkkan, N. (1992). Sample size determination in Bayesian analysis. *The Statistician* **41**, 389–397.

Piantadosi, S. (1997). *Clinical Trials: A Methodologic Perspective.* New York: Wiley.

Pliskin, J. S., Shepard, D., and Weinstein, M. C. (1980). Utility functions for life years and health status: Theory, assessment, and application. *Operations Research* **28**, 206–224.

Pryse-Phyllips, W. E. M., Dodick, D. W., Edmeads, J. G., Gawel, M. J., Nelson, R. F., Purdy, R. A., Robinson, G., Stirling, D., and Worthington, I. (1997). Guidelines for the diagnosis and management of migraine in clinical practice. *Canadian Medical Association Journal* **156**, 1273–1287.

Quiet, C., Ferguson, D., Weichelsbaum, R. R., and Hellman, S. (1996). Natural history of node-negative breast cancer: The curability of small cancers with a limited number of positive nodes. *Journal of Clinical Oncology* **14**, 3105–3111.

Raiffa, H. and Schleifer, R. (1961). *Applied Statistical Decision Theory.* Boston: Harvard University Press.

Ramsey, F. P. (1931). Truth and probability. In: F. P. Ramsey, *The Foundations of Mathematics and Other Logical Essays* (ed. R. B. Braithwaite). London: Kegan Panl, Trench, Trubner & Co.

Ramsey, F. P. (1990). Weight or the value of knowledge. *British Journal for the Philosophy of Science* **41**, 1–4.

Ramsey, S. D., McIntosh, M., Etzioni, R., and Urban, N. (2000). Simulation modeling of outcomes and cost effectiveness. *Hematology/Oncology Clinics of North America* **14**(4), 925–938.

Raudenbush, S. W. and Bryk, A. S. (1985). Empirical Bayes meta-analysis. *Journal of Educational and Behavioral Statistics* **10**, 75–98.

Recht, A. and Houlihan, M. J. (1995). Axillary lymph nodes and breast cancer: A review. *Cancer* **76**(9), 1491–1512.

Redelmeier, D. A. and Heller, D. N. (1993). Time preference in medical decision making and cost-effectiveness analysis. *Medical Decision Making* **13**, 212–217.

Ripley, B. D. (1987). *Stochastic Simulation.* New York: Wiley.

Roa, B. B., Boyd, A. A., Volcik, K., and Richards, C. S. (1996). Ashkenazi Jewish population frequencies for common mutations in *BRCA1* and *BRCA2*. *Nature Genetics* **14**, 185–187.

Robert, C. P. (1994). *The Bayesian Choice.* Berlin: Springer-Verlag.

Robert, C. P. and Casella, G. (1999). *Monte Carlo Statistical Methods.* New York: Springer-Verlag.

Rosen, P., Groshen, S., Saigo, P., Kinne, D., and Hellman, S. (1989). A long term follow-up study of survival in stage I ($T_1N_0M_0$) and stage II ($T_1N_1M_0$) breast carcinoma. *Journal of Clinical Oncology* **7**, 355–366.

Rosenberg, R. D., Hunt, W. C., Williamson, M. R., Gilliland, F. D., Wiest, P. W., Kelsey, C. A., Key, C. R., and Linver, M. N. (1998). Effects of age, breast density, ethnicity, and estrogen replacement therapy on screening mammographic sensitivity and cancer stage at diagnosis: Review of 183,134 screening mammograms in Albuquerque, New Mexico. *Radiology* **209**(2), 511–518.

Rubin, D. B. (1984). Bayesianly justifiable and relevant frequency calculations for the applied statistician. *Annals of Statistics* **12**, 1151–1172.

Rustagi, J. S. (1976). *Variational Methods in Statistics.* New York: Academic Press.

Sackett, D. L., Rosenberg, W. M. C., Gray, J. A. M., Haynes, R. B., and Richardson, W. S. (1996). Evidence based medicine: What it is and what it isn't. *British Medical Journal* **312**, 71–72.

Sacks, H. S., Berrier, J., Reitman, D., Ancona-Berk, V. A., and Chalmers, T. C. (1987). Meta-analyses of randomized controlled trials. *New England Journal of Medicine* **316**, 450–455.

Saltzmann, P., Kerlikowske, K., and Phillips, K. (1997). Cost-effectiveness of extending screening mammography guidelines to include women 40 to 49 years of age. *Annals of Internal Medicine* **127**, 955–965.

Samsa, G. P., Reutter, R. A., Parmigiani, G., Hasselblad, V., Lipscomb, J., and Matchar, D. B. (1999). Methods for validating complex simulation-based decision models: Applications to the Stroke Prevention Policy Model. *Journal of Clinical Epidemiology* **52**, 259–271.

Savage, L. J. (1954). The Foundations of Statistics. New York: Wiley.

Schervish, M. J. (1995). Theory of Statistics. New York: Springer-Verlag.

Schildkraut, J., Iversen, Jr, E. S., Parmigiani, G., and Berry, D. (1997). Prognostic significance of estimated *BRCA1* and *BRCA2* mutation status in women diagnosed with breast cancer. *Genetic Epidemiology* **14**, 538.

Schwartz, M. (1978). A mathematical model used to analyze breast cancer screening strategies. *Operations Research* **26**, 937–955.

Schwartz, M., Lerman, C., Daly, M., Audrain, J., Masny, A., and Griffith, K. (1995). Utilization of ovarian cancer screening by women at increased risk. *Cancer Epidemiology, Biomarkers and Prevention* **4**(3), 269–273.

Scott, G. C., Cher D. J., and Lenert, L. A. (1997). SecondOpinion: Interactive web-based access to a decision model. In: D. R. Masys (ed.), *Proceedings of the AMIA Annual Full Symposium*, pp. 769–773. Bethesda, MD: American Medical Informatics Association.

Scully, R., Chen, J., Plug, A., Xiao, Y., Weaver, D., Feunteun, J., Ashley, T., and Livingston, D. M. (1997). Association of *BRCA1* with *Rad51* in mitotic and meiotic cells. *Cell* **88**(2), 265–275.

Seidenfeld, T. (1988). Decision theory without 'independence' or without 'ordering'. What is the difference? *Economics and Philosophy* **4**, 267–290.

Seidenfeld, T., Schervish, M. J., and Kadane, J. B. (1995). A representation of partially ordered preferences. *Annals of Statistics* **23**, 2168–2217.

Sengupta, B. (1980). Inspection procedures when failure symptoms are delayed. *Operations Research* **28**, 768–776.

Shahani, A. K. and Crease, D. M. (1977). Towards models of screening for early detection of disease. *Advances in Applied Probability* **9**, 665–680.

Shapiro, S. (1997). Periodic screening for breast cancer: The hip randomized controlled trial. *Journal of the National Cancer Institute Monographs* **22**, 27–30.

Shapiro, S., Venet, W., Strax, P., and Venet, L. (1988). *Periodic Screening for Breast Cancer*. Baltimore, MD: Johns Hopkins University Press.

Shattuck-Eidens, D., Oliphant, A., McClure, M., McBride, C., Gupte, J., *et al.* (1997). *BRCA1* sequence analysis in women at high risk for susceptibility mutations. *Journal of the American Medical Association* **278**, 1242–1250.

Shoemaker, P. (1982). The expected utility model: Its variants, purposes, evidence and limitations. *Journal of Economic Literature* **20**, 529–563.

Shrout, P. E. and Newman, S. C. (1989). Design of two-phase prevalence surveys of rare disorders. *Biometrics* **45**, 549–555.

Siegel, C., Laska, E., and Meisner, M. (1996). Statistical methods for cost-effectiveness analyses. *Controlled Clinical Trials* **17**, 387–406.

Silliman, N. (1997). Hierarchical selection models with applications in meta-analysis. *Journal of the American Statistical Association* **92**, 926–936.

Silverstein, M. J., Gierson, E. D., Waisman, J. R., Senofsky, G. M., Colburn, W. J., and Gamagami, P. (1994). Axillary lymph node dissection for T1a breast carcinoma. Is it indicated? *Cancer* **73**(3), 664–667.

Simes, R. J. (1986). Publication bias: The case for an international registry of clinical trials. *Journal of Clinical Oncology* **4**, 1529–1541.

Sloan, F. (1995). Valuing Health Care. Cambridge: Cambridge University Press.

Smith, T. C., Spiegelhalter, D. J., and Thomas, A. (1995). Bayesian approaches to random-effects meta-analysis: A comparative study. *Statistics in Medicine* **14**, 2685–2699.

Smith, T. C., Spiegelhalter, D. J., and Parmar, M. K. B. (2000). Bayesian meta-analysis of randomized trials using graphical models and BUGS. In D. A. Berry and D. Stangl (eds), *Meta-analysis in Medicine and Health Policy*. New York: Marcel Dekker.

Smith, T. E., Harrold, E. V., Matloff, E. T., Turner, B. C., and Haffty, B. G. (1996). *BRCA1/2* status, family history and ipsilateral breast tumor relapse (IBTR) following conservative surgery and radiation therapy (CS + RT). *Proceedings of the American Society of Clinical Oncology* **15**, Abs # 2413.

Sorenson, R. M. and Pace, N. L. (1992). Anesthetic techniques during surgical repair of femoral neck fractures. A meta-analysis. *Anesthesiology* **77**, 1095–1104.

Sox, H. (1986). Probability theory in the use of diagnostic tests. An introduction to critical study of the literature. *Annals of Internal Medicine* **104**(1), 60–66.

Sox, H. C., Blatt, M. A., Higgins, M. C., and Marton, K. I. (1988). Medical Decision Making. Boston: Butterworth-Heinemann.

Spiegelhalter, D. J., Thomas, A., Best, N., and Gilks, W. R. (1996). *BUGS 0.5: Bayesian inference Using Gibbs Sampling*. MRC Biostatistics Unit, Cambridge.

Spratt, J. S., Greenberg, R. A., and Heuser, L. S. (1986). Geometry, growth rates, and duration of cancer and carcinoma in situ of the breast before detection by screening. *Cancer Research* **46**, 970–974.

Stangl, D. K. and Berry, D. A. (eds) (2000). Meta-analysis in Medicine and Health Policy. New York: Marcel Dekker.

Sterling, T. D., Rosenbaum, W. L., and Weinkam, J. J. (1995). Publication decisions revisited: The effect of the outcome of statistical tests on the decision to publish and vice versa. *American Statistician* **49**, 108–112.

Stewart, W. F., Lipton, R. B., Celentano, D. D., and Reed, M. L. (1992). Prevalence of migraine headache in the United States. Relation to age, income, race, and other sociodemographic factors. *Journal of the American Medical Association* **267**(1), 64–69.

Stigler, S. M. (1986). History of Statistics. Cambridge, MA: Harvard University Press.

Stinnett, A. A. and Mullahy, J. (1998). Net health benefits: A new framework for the analysis of uncertainty in cost-effectiveness analysis. *Medical Decision Making* **18**(2. Suppl).

Straatman, H., Peer, P. G. M., and Verbeek, A. L. (1997). Estimating lead time and sensitivity in a screening program without estimating the incidence in the screened group. *Biometrics* **53**, 217–229.

Stroup, D. F., Berlin, J. A., Morton, S. C., Olkin, I., Williamson, G. D., Rennie, D., Moher, D., Becker, B. J., Sipe, T. A., and Thacker, S. B. (2000). Meta-analysis of observational studies in epidemiology: a proposal for reporting. Meta-analysis of Observational Studies in Epidemiology (MOOSE) group. *Journal of the American Medical Association* **283**, 2008–2012.

Struewing, J. P., Hartge, P., Wacholder, S., Baker, S. M., Berlin, M., McAdams, M., Timmerman, M. M., Brody, L. C., and Tucker, M. A. (1997). The risk of cancer associated with specific mutations of *BRCA1* and *BRCA2* among Ashkenazi Jews. *New England Journal of Medicine* **336**, 1401–1408.

Sutton, A. J., Abrams, K. R., Jones, D. R., Sheldon, T. A., and Song, F. (1998). Systematic reviews of trials and other studies. *Health Technology Assessment* **2**(19).

Sutton, A. J., Lambert, P. C., Hellmich, M., Abrams, K. R., and Jones, D. R. (2000). Meta-analysis in practice: A critical review of available software. In: D. K. Stangl and D. A. Berry (eds), Meta-analysis in Medicine and Health Policy, pp. 359–390. New York: Marcel Dekker

Szolovits, P. (1995). Uncertainty and decisions in medical informatics. *Methods of Information in Medicine* **34**, 111–121.

Szolovits, P. and Pauker, S. (1992). Pedigree analysis for genetic counseling. In: K. C. Lun, P. Degoulet, T. E. Piemme, and O. Rienhoff (eds), MEDINFO-92: *Proceedings of the Seventh Conference on Medical Informatics*, pp. 679–683, New York: Elsevier.

Tabár, L., Chen, H.-H., Fagerberg, G., Duffy, S., and Smith, T. (1997). Recent results from the Swedish two-county trial: The effects of age, histologic type and mode of detection on the efficacy of breast cancer screening. *Journal of the National Cancer Institute Monographs* **22**, 43–47.

Tan, S. and Smith, A. (1998). Exploratory thoughts on clinical trials with utilities. *Statistics in Medicine* **17**, 2771–2791.

Tanner, M. A. (1991). *Tools for Statistical Inference – Observed Data and Data Augmentation Methods*, Lecture Notes in Statistics 67. New York: Springer-Verlag.

Tanner, M. A. and Wong, W. H. (1987). The calculation of posterior distributions by data augmentation. *Journal of the American Statistical Association* **82**, 528–550.

Thernau, T. M. (1996). *A Package for Survival Analysis in S*. Rochester, MN: Mayo Foundation.

Thompson, S. G. (1994). Systematic review: Why sources of heterogeneity in meta-analysis should be investigated. *British Medical Journal* **309**, 1351–1355.

Tierney, L. (1994). Markov chains for exploring posterior distributions (with discussion). *Annals of Statistics* **22**(4), 1701–1762.

Torrance, G. W., Thomas, W., and Sackett, D. (1972). A utility maximization model for evaluation of health care programs. *Health Services Research* **7**, 118–133.

Torrance, G. W. (1986). Measurement of health state utilities for economic appraisal. *Journal of Health Economics* **5**, 1–30.

Tsodikov, A. D., Asselain, B., Fourque, A., Hoang, T., and Yakovlev, A. Y. (1995). Discrete strategies of cancer post-treatment surveillance. Estimation and optimization problems. *Biometrics* **51**, 437–447.

Tsodikov, A. D. and Yakovlev, A. Y. (1991). On the optimal policies of cancer screening. *Mathematical Biosciences* **107**, 21–45.

Tweedie, R. L., Scott, D. J., Biggerstaff, B. J., and Mengersen, K. L. (1996). Bayesian meta-analysis, with application to studies of ETS and lung cancer. *Lung Cancer* **14**(1 Suppl.), 171–194.

van Ineveld, B. M., van Oortmarssen, G. J., de Koning, H. J., Boer, R., and van der Maas, P. J. (1993). How cost-effective is breast cancer screening in different EC countries? *European Journal of Cancer* **29**, 1663–1668.

van Meerten, R. J., Durinick, J. R., and Dewit, C. (1971). Computer guided diagnosis of asthma, asthmatic bronchitis, chronic bronchitis and amphysema: Computing, methods and results. *Respiration* **28**, 399–408.

van Oortmarseen, G., Boer, R., and Habbema, J. (1995). Modeling issues in cancer screening. *Statistical Methods in Medical Research* **4**, 33–54.

Velanovich, V. (1998). Axillary lymph node dissection for breast cancer: A decision analysis of T1 lesions. *Annals of Surgical Oncology* **5**(2), 131–139.

Vernon, S. W. (1999). Risk perception and risk communication for cancer screening behaviors: A review. *Journal of the National Cancer Institute Monographs* **25**, 101–119.

Vitak, B., Olsen, K. E., Manson, J. C., Arnesson, L. G., and Stal, O. (1999). Tumour characteristics and survival in patients with invasive interval breast cancer classified according to mammographic findings at the latest screening: A comparison of true interval and missed interval cancers. *European Journal of Radiology* **9**, 460–469.

von Neumann, J. and Morgenstern, O. (1944). *Theory of Games and Economic Behavior*. New York: Wiley.

Wald, A. (1949). *Statistical Decision Functions*. New York: Wiley.

Walter, S. D. and Day, N. E. (1983). Estimation of the duration of a pre-clinical disease state using screening data. *American Journal of Epidemiology* **118**, 865–886.

Warmuth, M. A., Bowen, G., Prosnitz, L. R., Chu, L., Broadwater, G., Peterson, B., Leight, G., and Winer, E. P. (1998). Complications of axillary lymph node dissection for carcinoma of the breast: A report based on a patient survey. *Cancer* **83**(7), 1362–1368.

Warmuth, M. A., Sutton, L. M., and Winer, E. P. (1997). A review of hereditary breast cancer risk: From screening to risk factor modification. *American Journal of Medicine* **102**, 407–415.

Weber, B. (1996). Genetic testing for breast cancer. *Science & Medicine* **3**, 12–21.

Weber, B. L. (1998). Update on breast cancer susceptibility genes. *Recent Results in Cancer Research* **152**, 49–59.

Weinstein, M. C. (1989). Methodologic issues in policy modeling for cardiovascular disease. *Journal of the American College of Cardiology* **14** (3 Suppl. A), 38A–43A.

Weinstein, M. C., Coxson, P. G., Williams, L. W., Pass, T. M., Stason, W. B., and Goldman, L. (1987). Forecasting coronary heart disease incidence, mortality, and cost: The coronary heart disease policy model. *American Journal of Public Health* **77**, 1417–1426.

Weinstein, M. C., Feinberg, H., Elstein, A. S., Frazier, H. S., Neuhauser, D., Neutra, R. R., and McNeil, B. J. (1980). *Clinical Decision Analysis*. Philadelphia: Saunders.

Weinstein, M. C. and Stason, W. B. (1977). Foundations of cost-effectiveness analysis for health and medical practices. *New England Journal of Medicine* **296**, 716–721.

Weiss, G. H. and Lincoln, T. L. (1966). Analysis of repeated examinations for the detection of occult disease. *Health Services Research* **1**, 272–286.

Weiss, K. M. (1993). *Genetic Variation and Human Disease*. Cambridge: Cambridge University Press.

West, M. and Turner, D. (1994). Deconvolution of mixtures in analysis of neural synaptic transmission. *The Statistician* **43**, 31–43.

Wiener, D. A., Ryan, T. J., McCabe, C. H., Kennedy, J. W., Schloss, M., Tristani, F., Chaitman, B. R., and Fisher, L. D. (1979). Exercise stress testing. Correlations among history of angina, ST-segment response and prevalence of coronary-artery disease in the Coronary Artery Surgery Study (CASS). *New England Journal of Medicine* **301**, 230–235.

Wood, W., Weiss, R., Tormey, D., Holland, J., Henry, P., Leone, L., *et al.* (1985). A randomized trial of CMF versus CMFVP as adjuvant chemotherapy in women with node-positive stage II breast cancer: A CALGB study. *World Journal of Surgery* **9**, 714–718.

Wood, W., Budman, D., Korzun, A., Cooper, M., Younger, J., Hart, R., *et al.* (1994). Dose and dose intensity of adjuvant chemotherapy for stage II, node-positive breast carcinoma. *New England Journal of Medicine* **330**, 1253–1259.

Wooster, R., Bignell, G., Lancaster, J., Swift, S., Seal, S., Mangion, J., Collins, N, Gregory, S., Gumbs, C., and Miklem, G. (1995). Identification of the breast cancer susceptibility gene *BRCA2*. *Nature* **378**, 789–92.

Yakovlev, A. Y. and Tsodikov, A. D. (1996). *Stochastic Models of Tumor Latency and Their Biostatistical Applications*. Singapore: World Scientific.

Yang, R. and Berger, J. O. (1997). A catalogue of noninformative priors. Discussion paper 97-42, Institute of Statistics and Decision Sciences, Duke University.

Yao, T.-J., Begg, C. B., and Livingston, P. O. (1996). Optimal sample size for a series of pilot trials of new agents. *Biometrics* **52**, 992–1001.

Zelen, M. (1976). The theory of early detection of breast cancer in the general population. In J. C. Heuson, W. H. Mattheiem, and M. Rozencweig (eds), Breast Cancer: Trends in Research and Treatment, pp. 287–300. New York. Raven Press.

Zelen, M. (1993). Optimal schedules of examinations for early detection of disease. *Biometrika* **80**, 279–294.

Zelen, M. and Feinleib, M. (1969). On the theory of screening for chronic diseases. *Biometrika* **56**, 601–614.

Ziegler, D. K. (1990). Headache. Public health problem. *Neurol. Clin.* **8**, 781–791.

Index

Page numbers in italic, e.g. *14*, signify references to figures. Page numbers in bold, e.g. **13**, denote references to tables.

STATISTICS IN PRACTICE

Human and Biological Sciences

Brown and Prescott – Applied Mixed Models in Medicine
Marubini and Valsecchi – Analysing Survival Data from Clinical Trials and
Observation Studies
Parmigiani – Modeling in Medical Decision Making: A Bayesian Approach
Senn – Cross-over Trials in Clinical Research
Senn – Statistical Issues in Drug Development
Whitehead – Design and Analysis of Sequential Clinical Trials, Revised Second
Edition

Earth and Environmental Sciences

Buck, Cavanagh and Litton – Bayesian Approach to Intrepreting Archaeolog-
ical Data
Webster and Oliver – Geostatistics for Environmental Scientists

Industry, Commerce and Finance

Aitken – Statistics and the Evaluation of Evidence for Forensic Scientists
Lehtonen and Pahkinen – Practical Methods for Design and Analysis of
Complex Surveys
Ohser and Mücklich – Statistical Analysis of Microstructures in Materials
Science